The Open University

M381 Number Theory and
Mathematical Logic

GW00598018

Number Theory **Unit 5**

Multiplicative Functions

Prepared for the Course Team by Alan Best

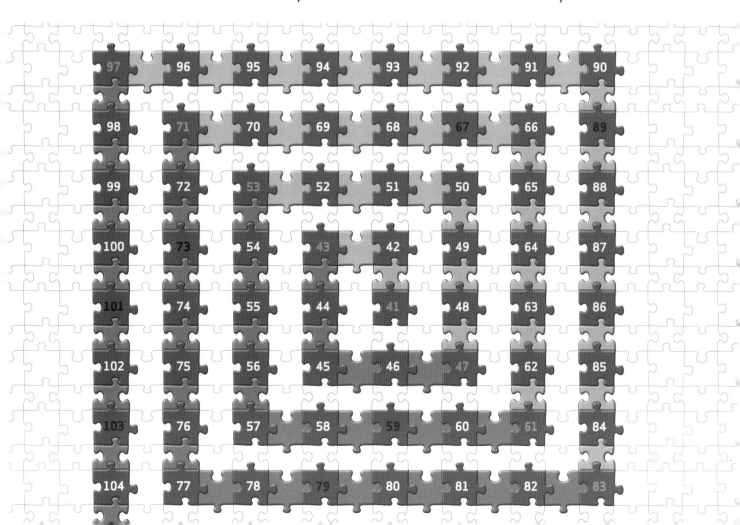

The M381 Number Theory Course Team

The Number Theory half of the course was produced by the following team:

Alan Best	*Author*
Andrew Brown	*Course Team Chair* and *Academic Editor*
Roberta Cheriyan	*Course Manager*
Bob Coates	*Critical Reader*
Dick Crabbe	*Publishing Editor*
Janis Gilbert	*Graphic Artist*
Derek Goldrei	*Critical Reader*
Caroline Husher	*Graphic Designer*
John Taylor	*Graphic Artist*

with valuable assistance from:

CMPU	*Mathematics and Computing, Course Materials Production Unit*
John Bayliss	*Reader*
Elizabeth Best	*Reader*
Jeremy Gray	*History Reader*
Alison Neil	*Reader*

The external assessor was:

Alex Wilkie	*Reader in Mathematical Logic, University of Oxford*

This publication forms part of an Open University course. Details of this and other Open University courses can be obtained from the Student Registration and Enquiry Service, The Open University, PO Box 197, Milton Keynes, MK7 6BJ, United Kingdom: tel. +44 (0)870 300 6090, e-mail general-enquiries@open.ac.uk

Alternatively, you may visit the Open University website at http://www.open.ac.uk where you can learn more about the wide range of courses and packs offered at all levels by The Open University.

To purchase a selection of Open University course materials, visit http://www.ouw.co.uk, or contact Open University Worldwide, Michael Young Building, Walton Hall, Milton Keynes, MK7 6AA, United Kingdom, for a brochure: tel. +44 (0)1908 858793, fax +44 (0)1908 858787, e-mail ouw-customer-services@open.ac.uk

The Open University, Walton Hall, Milton Keynes, MK7 6AA.

First published 1996. Reprinted 1997 and 2001. New edition 2007 with corrections.

Edited, designed and typeset by The Open University, using the Open University TeX System.

Printed and bound in the United Kingdom by The Charlesworth Group, Wakefield.

ISBN 978 0 7492 2275 8

2.1

CONTENTS

INTRODUCTION

From the time of Fermat's announcement of FLT, 120 years passed before a proof was published. That proof was given by Euler who, some 24 years later, went on to prove a generalization of FLT which caters for the case of the modulus being composite. In doing so he introduced a function which we nowadays refer to as Euler's ϕ-*function*; it is a function whose importance in number theory cannot be overestimated. The ϕ-function, its properties and the generalization of FLT form the core of this unit.

Euler is pronounced *oiler*.

ϕ is the Greek letter phi.

In this unit we shall be much concerned with *number-theoretic functions*, that is, integer-valued functions defined on the set of positive integers. When a number-theoretic function f has the property that $f(mn) = f(m)f(n)$, whenever $\gcd(m, n) = 1$ it is called a *multiplicative* function. Multiplicative functions will be found to have the desirable property that their values everywhere are determined by their values at prime powers. We shall meet numerous examples of multiplicative functions in this unit, the ϕ-function being one of them. We shall begin by investigating two other classical multiplicative functions: the τ function which counts the number of divisors of a positive integer, which we have already encountered in *Unit 2*, and the σ function which adds up the divisors of a positive integer.

The number 6 has the interesting property that it is equal to the sum of all its divisors excluding itself.

$$6 = 1 + 2 + 3.$$

Numbers with this property are called *perfect* numbers and, as we shall report in this unit, not many have been found. The search for perfect numbers has a fascinating history, having occupied mathematical minds since the time of the early Greeks. This search is still pursued actively today. When investigating perfect numbers we shall discover the so-called *Mersenne* primes; a family of rare primes each of which yields a perfect number.

In *Unit 4* we touched upon the ideas of order and primitive roots when working to some prime modulus. With the benefit of the ϕ-function we shall look again at those notions, generalizing and extending the earlier work to cater for a composite modulus.

Throughout this unit one name will keep appearing: that of Euler.

Leonhard Euler (1707–1783)

Whilst Fermat's work on number theory had revitalized the subject, it was disappointing that he failed to interest many other distinguished mathematicians of his time in his work on numbers. He died unaware of the influence his endeavours were to have on future generations of number theorists. The first to really appreciate Fermat's work was Leonhard Euler.

Euler was born in Basel, Switzerland. At the age of 13 he was sent to the University of Basel to study theology so that he might become a clergyman, like his father. However, his father had also been a student of mathematics and had imparted some of his interest in this area to Leonhard. At university Euler met Johann Bernoulli, and befriended him and his two sons Nicolas and Daniel, three of Switzerland's finest mathematicians. Euler soon abandoned his studies in theology to concentrate on mathematics, and in no time he began to establish an international reputation as a mathematician. He is nowadays regarded as the greatest mathematician of the 18th century.

In 1727 Euler secured a position, nominally in natural philosophy, at the recently-founded Imperial Academy at St. Petersburg, where the young Bernoulli brothers had gone as professors of mathematics. He managed to pursue his mathematical interests there and in 1733 succeeded the

Bernoullis to the chair of mathematics when Daniel returned to Basel, Nicolas having been tragically killed previously in an accident. Euler remained at St. Petersburg until 1741 and subsequently returned there in 1766 having spent the intervening years at the Royal Academy in Berlin.

Euler's output was incredible. At the age of 18 he published his first paper (on a design for the arrangement of masts on a ship), and he went on to publish over 500 books and papers during his lifetime. After his death, it took the St. Petersburg Academy 47 years to publish all the manuscripts he had left behind. His writings covered the complete spectrum of mathematics including applications of calculus, algebra, co-ordinate and differential geometry, infinite series and number theory.

In 1735 Euler lost the use of his right eye; through overwork it is said. Early in his second period in Russia, Euler began to lose sight in his left eye and, to prepare for the impending blindness, he began practising dictating mathematics and developing his already phenomenal memory; it is said that he knew Virgil's Aeneid by heart. He spent the last seventeen years of his life able to read only large letters chalked on a blackboard; but this did not stop his researches. For example, in 1772 he dictated an 800 page book on lunar motion and in 1776 alone he had 50 research papers published.

Amongst his many outstanding achievements Euler will be remembered for the way he introduced symbolism and notation into mathematics. He was the first to adopt the symbol e for the base of the natural logarithms and i for a (complex) square-root of -1. These two symbols for which Euler is responsible are combined with π, whose use he helped to establish, in a famous equation which bears his name, $e^{i\pi} + 1 = 0$.

As far as number theory is concerned, we have already mentioned that it was Euler who came up with the first published proof of FLT. In this and subsequent units we shall witness many more of Euler's contributions to the subject.

1 MULTIPLICATIVE FUNCTIONS

1.1 The τ and σ functions

In *Unit 2* we met the τ function. For convenience we shall recall the definition here.

Definition 1.1 The τ function

For any integer $n \geq 1$, $\tau(n)$ is defined to be the number of distinct divisors of n, including 1 and n.

The τ function is an example of an integer-valued function which has the set of positive integers, $\mathbb{Z}^+$, as its domain. We shall refer to such functions $f : \mathbb{Z}^+ \longrightarrow \mathbb{Z}$ as being *number-theoretic* functions. Although a number-theoretic function is permitted to take any integer value, in all the examples that we shall meet in this course the values taken will be positive integers or 0.

Number-theoretic functions are also called arithmetic functions.

Closely related to the τ function is a second number-theoretic function denoted by σ.

σ is the Greek letter sigma.

Definition 1.2 The σ function

For any integer $n \geq 1$, $\sigma(n)$ is defined to be the sum of the distinct divisors of n, including 1 and n.

So, for example, as the divisors of 20 are 1, 2, 4, 5, 10 and 20 we have

$$\sigma(20) = 1 + 2 + 4 + 5 + 10 + 20 = 42.$$

Problem 1.1 _____

Complete the following table:

n	1	2	3	4	5	6	7	8	9	10
$\tau(n)$										
$\sigma(n)$										

Before examining these two functions in more detail let us introduce some notation which will be of assistance in a number of situations. The Greek letter capital sigma is customarily used to denote summation. We extend that notation and write

$$\sum_{d|n} f(d)$$

to denote the sum of terms $f(d)$ taken over all positive divisors d of n. For example,

$$\sum_{d|10} f(d) = f(1) + f(2) + f(5) + f(10).$$

With this notation we observe the following.

Divisor sum formulae for τ and σ

$$\tau(n) = \sum_{d|n} 1 \quad \text{and} \quad \sigma(n) = \sum_{d|n} d$$

In addition to summation, we shall be concerned in this unit with products and we have a corresponding notation in which sigma is replaced by the Greek letter capital pi. The following examples illustrate typical uses.

$$\prod_{1 \leq d \leq 4} f(d) = f(1)f(2)f(3)f(4)$$

$$\prod_{\substack{p|28 \\ p \text{ prime}}} f(p) = f(2)f(7)$$

In the second of these examples the product is taken over all prime divisors of 28.

Problem 1.2 _____

Determine $\displaystyle\sum_{d|8} \sigma(d)$ and $\displaystyle\prod_{d|8} \tau(d)$.

Problem 1.3 _____

Show that

$$\sum_{d|n} f(d) = \sum_{d|n} f\left(\frac{n}{d}\right).$$

Hint: If d is a divisor of n then $\frac{n}{d}$ is also a divisor of n.

Deduce from this formula that

$$\sum_{d|n} \frac{1}{d} = \frac{\sigma(n)}{n}.$$

It is a simple matter to determine the values taken by each of the τ and σ functions at primes. If p is a prime then, as the only divisors of p are itself and 1, $\tau(p) = 2$ and $\sigma(p) = p + 1$. Indeed an integer $n > 1$ is prime if, and only if, $\tau(n) = 2$, or equivalently, if, and only if, $\sigma(n) = n + 1$, for each of these statements amounts to the fact that the only divisors of n are itself and 1.

If n is composite then $\tau(n) > 2$ and $\sigma(n) > n + 1$.

We can readily go a stage further and look at powers of the prime p. As the divisors of p^k are 1, p, p^2, ..., p^{k-1} and p^k, we have

$$\tau\left(p^k\right) = k + 1 \quad \text{and} \quad \sigma\left(p^k\right) = 1 + p + p^2 + \cdots + p^k = \frac{p^{k+1} - 1}{p - 1}.$$

The second expression for $\sigma(p^k)$ utilizes the formula for the sum of a geometric series.

In *Unit 2* we completed the task for the τ function obtaining the following formula for the value of $\tau(n)$ at any positive integer n given in its prime decomposition.

$$\tau\left(p_1^{k_1} p_2^{k_2} \ldots p_r^{k_r}\right) = (k_1 + 1)(k_2 + 1) \ldots (k_r + 1).$$

We may refer to this as the prime decomposition formula for τ.

But notice that as $\tau\left(p_i^{k_i}\right) = (k_i + 1)$ we have from this the following formula.

$$\tau\left(p_1^{k_1} p_2^{k_2} \ldots p_r^{k_r}\right) = \tau\left(p_1^{k_1}\right) \tau\left(p_2^{k_2}\right) \ldots \tau\left(p_r^{k_r}\right)$$

This shows that the value of the τ function at any integer n is completely determined by its values at the prime power divisors of n. In other words, we know all about τ once we know its value at a general prime power. Functions with this property command considerable interest in number theory.

1.2 Multiplicative functions

Definition 1.3 Multiplicative functions

A number-theoretic function is said to be *multiplicative* if, for any integer with prime decomposition $p_1^{k_1} p_2^{k_2} \ldots p_r^{k_r}$,

$$f\left(p_1^{k_1} p_2^{k_2} \ldots p_r^{k_r}\right) = f\left(p_1^{k_1}\right) f\left(p_2^{k_2}\right) \ldots f\left(p_r^{k_r}\right).$$

So τ has provided us with our first example of a multiplicative function, and it will come as no surprise to learn that the σ function will provide a second example. Before establishing that fact let us practice making use of the multiplicative nature of such functions.

Problem 1.4

In this problem you may assume the result that τ and σ are multiplicative functions.

(a) Determine $\tau(360)$ and $\sigma(360)$.

(b) If $p > 3$ is prime, determine $\tau(2p^3)$, $\sigma(3p)$ and $\sigma(6p^2)$.

The ability to calculate values of multiplicative functions at composite numbers in the way illustrated in Problem 1.4 will prove to be very useful. However, there is an alternative way of viewing the definition of a multiplicative function which will simplify matters in some situations.

Theorem 1.1 Equivalent definition of multiplicative functions

The function f is a multiplicative function if, and only if, for every pair, m and n, of relatively prime positive integers,

$$f(mn) = f(m)f(n).$$

Proof of Theorem 1.1

Suppose first that f is multiplicative, and let $m = p_1^{k_1} p_2^{k_2} \ldots p_r^{k_r}$ and $n = q_1^{l_1} q_2^{l_2} \ldots q_s^{l_s}$ be relatively prime integers. As m and n are relatively prime, the primes p_i and q_j are distinct and so

$$f(mn) = f\left(p_1^{k_1}\right) f\left(p_2^{k_2}\right) \ldots f\left(p_r^{k_r}\right) f\left(q_1^{l_1}\right) f\left(q_2^{l_2}\right) \ldots f\left(q_s^{l_s}\right) = f(m)f(n).$$

Conversely, suppose that $f(mn) = f(m)f(n)$ whenever $\gcd(m,n) = 1$. Then, for any integer $p_1^{k_1} p_2^{k_2} \ldots p_r^{k_r}$ in prime decomposition,

$$f\left(p_1^{k_1} p_2^{k_2} \ldots p_r^{k_r}\right) = f\left(p_1^{k_1}\right) f\left(p_2^{k_2} \ldots p_r^{k_r}\right),$$

$$\text{since } \gcd\left(p_1^{k_1}, p_2^{k_2} \ldots p_r^{k_r}\right) = 1,$$

$$= f\left(p_1^{k_1}\right) f\left(p_2^{k_2}\right) f\left(p_3^{k_3} \ldots p_r^{k_r}\right),$$

$$\text{since } \gcd\left(p_2^{k_2}, p_3^{k_3} \ldots p_r^{k_r}\right) = 1,$$

$$= \ldots$$

$$= f\left(p_1^{k_1}\right) f\left(p_2^{k_2}\right) f\left(p_3^{k_3}\right) \ldots f\left(p_r^{k_r}\right),$$

The argument can be made watertight by the use of Mathematical Induction.

and so f is multiplicative. ∎

Problem 1.5

If f is a multiplicative function prove that, with the exception of one function, $f(1) = 1$. What is the exceptional multiplicative function for which this does not hold?

The facts that σ and τ are multiplicative functions will fall out as consequences of the following general result.

Theorem 1.2 Generating new number-theoretic functions

If f is a multiplicative function, then the number-theoretic function F defined by

$$F(n) = \sum_{d|n} f(d)$$

is also multiplicative.

Proof of Theorem 1.2

Let m and n be relatively prime integers. Then

$$F(mn) = \sum_{d \mid mn} f(d).$$

Now the divisors of mn are the numbers of the form $d = rs$, where r and s are divisors of m and n respectively. Therefore we can write

$$F(mn) = \sum_{r \mid m, s \mid n} f(rs),$$

and so, appealing to the multiplicative property of f, we have

$$F(mn) = \sum_{r \mid m, s \mid n} f(r)f(s).$$

Example 4.2 of *Unit 1*.

Note that $\gcd(r, s) = 1$ since any common divisor of r and s would be a common divisor of the relatively prime integers m and n.

On the other hand

$$F(m)F(n) = \left(\sum_{r \mid m} f(r) \right) \left(\sum_{s \mid n} f(s) \right)$$

and, on multiplying out the product on the right, the sums for $F(mn)$ and $F(m)F(n)$ are seen to be the same. ∎

The final step of the above proof, noting that the product of two sums is equal to the appropriate sum of individual products, may need amplification. To clarify this, we shall illustrate the equality of the summations by tracing the above proof for a particular case.

Take $m = 4$ and $n = 25$. We get $F(4 \times 25)$, by summing the values of f over the 9 divisors of 100:

$$\begin{aligned} F(100) = {} & f(1) + f(2) + f(4) + f(5) + f(10) \\ & + f(20) + f(25) + f(50) + f(100). \end{aligned}$$

Next we write each of the nine divisors in its unique form as a divisor of 4 multiplied by a divisor of 25,

$$\begin{aligned} F(100) = {} & f(1 \times 1) + f(2 \times 1) + f(4 \times 1) + f(1 \times 5) + f(2 \times 5) \\ & + f(4 \times 5) + f(1 \times 25) + f(2 \times 25) + f(4 \times 25), \end{aligned}$$

and, appealing to the multiplicative property,

$$\begin{aligned} F(100) = {} & f(1)f(1) + f(2)f(1) + f(4)f(1) + f(1)f(5) + f(2)f(5) \\ & + f(4)f(5) + f(1)f(25) + f(2)f(25) + f(4)f(25). \end{aligned}$$

On the other hand

$$F(4) = f(1) + f(2) + f(4) \quad \text{and} \quad F(25) = f(1) + f(5) + f(25),$$

and so

$$F(4)F(25) = (f(1) + f(2) + f(4))(f(1) + f(5) + f(25)).$$

When this is multiplied out we get a sum of nine terms which can be seen to be precisely the nine terms in the final expression for $F(100)$.

So far, the number of examples of multiplicative functions that we have met is very limited. There are two more functions which we can quickly add to our list. The constant function c and the identity function id, which are defined by

$$c(n) = 1 \quad \text{and} \quad \text{id}(n) = n, \quad \text{for all } n \in \mathbb{Z},$$

are multiplicative because

$$c(mn) = 1 = c(m)c(n)$$

and

$$\text{id}(mn) = mn = \text{id}(m)\,\text{id}(n).$$

Note that for each of these functions the multiplicative property holds for all integers m and n, regardless of whether or not they are relatively prime.

9

Now we can apply Theorem 1.2, with the role of f being played in turn by the functions c and id, in the knowledge that each will give rise to a multiplicative function:

$$\sum_{d|n} c(d) = \sum_{d|n} 1 = \tau(n)$$

and

$$\sum_{d|n} \mathrm{id}(d) = \sum_{d|n} d = \sigma(n).$$

So we have proved the following.

Corollary to Theorem 1.2

The functions τ and σ are multiplicative.

Recalling that

$$\sigma(p^k) = (1 + p + p^2 + \cdots + p^k) = \frac{p^{k+1} - 1}{p - 1},$$

we can now use the multiplicative property to write down a formula for the value of $\sigma(n)$ in terms of the prime decomposition of n.

Theorem 1.3 The prime decomposition formula for σ

$$\sigma\left(p_1^{k_1} p_2^{k_2} \ldots p_r^{k_r}\right) = \left(1 + p_1 + p_1^2 + \cdots + p_1^{k_1}\right) \ldots \left(1 + p_r + p_r^2 + \cdots + p_r^{k_r}\right)$$

$$= \prod_{1 \leq i \leq r} \frac{p_i^{k_i + 1} - 1}{p_i - 1}$$

For example,

$$\sigma\left(2^4 \times 5^2 \times 7\right) = \frac{2^5 - 1}{2 - 1} \times \frac{5^3 - 1}{5 - 1} \times \frac{7^2 - 1}{7 - 1}$$

$$= \frac{31}{1} \times \frac{124}{4} \times \frac{48}{6}$$

$$= 31 \times 31 \times 8 = 7688.$$

Problem 1.6 _____

Show that the function $f : \mathbb{Z}^+ \longrightarrow \mathbb{Z}$ given by

$$f(n) = \begin{cases} 0, & \text{if } n \text{ is even,} \\ n, & \text{if } n \text{ is odd,} \end{cases}$$

is multiplicative.

Describe the multiplicative function $F(n) = \sum_{d|n} f(d)$.

Problem 1.7 _____

Prove that, for every positive integer r, the function $\sigma_r : \mathbb{Z}^+ \longrightarrow \mathbb{Z}^+$ defined by

$$\sigma_r(n) = \text{the sum of the } r\text{th powers of the positive divisors of } n$$

$$= \sum_{d|n} d^r,$$

is multiplicative.   *Hint*: Apply Theorem 1.2 to a suitably chosen function.

For example,

$$\sigma_4(6) = 1^4 + 2^4 + 3^4 + 6^4.$$

---

2 PERFECT NUMBERS

2.1 Even perfect numbers

A number is said to be *perfect* if it is equal to the sum of its parts (its positive divisors excluding itself). The first two perfect numbers are $6 = 1 + 2 + 3$ and $28 = 1 + 2 + 4 + 7 + 14$. Perfect numbers can be defined in terms of the σ function as follows.

Definition 2.1 Perfect numbers

A positive integer n is said to be *perfect* if it satisfies $\sigma(n) = 2n$.

As mentioned earlier in connection with the Pythagoreans, throughout history mathematicians and philosophers have continually attached religious and mystical significance to the properties of numbers. Pride of place in this respect goes to the perfect numbers. The perfection of the creation is embodied in the facts that, according to the Bible, God took 6 days to create all things, including the 28 day lunar cycle.

The search for perfect numbers has continued unabated for thousands of years. Only four were known to the ancient Greeks, namely 6, 28, 496 and 8128. It is perhaps not surprising that they had failed to extend this list because the next two perfect numbers turn out to be 33 550 336 and 8 589 869 056. However, Euclid made a significant contribution in Book IX of his *Elements* when he revealed the following result which we invite you to try your hand at proving.

Problem 2.1 ———————————————————————

Prove that, if k is a positive integer such that $2^k - 1$ is prime, then $n = 2^{k-1}(2^k - 1)$ is perfect.

———————————————————————

Some 2000 years after Euclid had published his result, Euler proved that the perfect numbers characterized by Euclid are the only *even* ones.

Theorem 2.1 Classification of even perfect numbers

If k is a positive integer such that $2^k - 1$ is prime, then $n = 2^{k-1}(2^k - 1)$ is perfect. Furthermore, every even perfect number is of this form for some positive integer k.

Proof of Theorem 2.1

We have proved that such a number n is perfect and it remains to prove that, conversely, every even perfect number must be of this given form.

Let n be an even perfect number, then $n = 2^r m$, where r is a positive integer and m is an odd positive integer. We can dismiss the case $m = 1$ because $\sigma(2^r) = 2^{r+1} - 1$ and this is not equal to 2×2^r. So we may assume that $m \geq 3$. Our goal is to show that $m = 2^{r+1} - 1$ and that it is prime.

Since 1 and m are distinct divisors of m we can write $\sigma(m) = m + t$, where $t \geq 1$. Now,

$$\sigma(n) = \sigma(2^r m) = \sigma(2^r)\sigma(m) = (2^{r+1} - 1)(m + t)$$
$$= 2^{r+1}m - m + (2^{r+1} - 1)t$$

and

$$2n = 2^{r+1}m.$$

As n is perfect, $\sigma(n) = 2n$, and equating the above expressions for $\sigma(n)$ and $2n$ leads to

$$m = \left(2^{r+1} - 1\right)t.$$

Hence t is a divisor of m and $t < m$ as 2^{r+1} is at least 4. Now if $t \neq 1$ then 1, t and m are three distinct divisors of m, so that $\sigma(m) \geq m + t + 1$, which contradicts $\sigma(m) = m + t$. We are forced to conclude that $t = 1$, which gives $m = 2^{r+1} - 1$. Moreover $\sigma(m) = m + 1$, which confirms that m is prime. Therefore $n = 2^r \left(2^{r+1} - 1\right)$ and $2^{r+1} - 1$ is prime. This shows that n is of the correct form with $r = k - 1$ and completes the proof. ∎

It may appear that Theorem 2.1 has solved the problem of even perfect numbers. We know exactly where to find them all; or do we?

2.2 Mersenne primes

Theorem 2.1 leaves one burning question. For which integers k is $2^k - 1$ prime? Each prime of this form leads to a perfect number and all even perfect numbers are obtained in this way. The list of candidates is quickly whittled down when we observe the identity

$$2^{rs} - 1 \equiv (2^r - 1)\left(2^{r(s-1)} + 2^{r(s-2)} + 2^{r(s-3)} + \cdots + 2^{2r} + 2^r + 1\right),$$

which shows that $2^k - 1$ is composite whenever $k = rs$ is composite. The question therefore becomes, for which primes p is $2^p - 1$ prime? For the first four primes, the value of $2^p - 1$ turns out to be prime, giving the first four perfect numbers:

Table 2.1 Four perfect numbers

p	$2^p - 1$	$2^{p-1}(2^p - 1)$
2	3	$2 \times 3 = 6$
3	7	$4 \times 7 = 28$
5	31	$16 \times 31 = 496$
7	127	$64 \times 127 = 8128$

Early writers believed that $2^p - 1$ is prime for all primes p, but this was eventually contradicted in 1536 when the factorization for the fifth number in this sequence was discovered:

$$2^{11} - 1 = 2047 = 23 \times 89.$$

By 1588 Cataldi, making use of a table of primes and testing for division by each prime less than the square root, had confirmed that $2^{13} - 1$, $2^{17} - 1$ and $2^{19} - 1$ are primes. He went on to claim that $2^p - 1$ is prime for the next four prime values of p, namely 23, 29, 31 and 37. However Fermat proved, around 1640, that two of these assertions were false by establishing that 47 divides $2^{23} - 1$ and that 223 divides $2^{37} - 1$.

Definition 2.2 Mersenne numbers

The numbers $M_p = 2^p - 1$, where p is prime, are known as *Mersenne numbers*. When M_p is prime it is referred to as a *Mersenne prime*.

Marin Mersenne (1588–1648)

Mersenne was a Franciscan friar, spending the main portion of his life in the Minim convent in Paris. From there he corresponded and became acquainted with many of the important mathematicians of that period, including Fermat. Indeed Mersenne was regarded as a 'centre for correspondence' for mathematics at that time. In 1644 Mersenne published the work for which he is best known, *Cogitata Physico-Mathematica*. In the preface to this book he made the claim that the only primes $p \leq 257$ for which $M_p = 2^p - 1$ is also prime are 2, 3, 5, 7, 13, 17, 19, 31, 67, 127 and 257. The last of these, 257, is the 55th prime and so this statement includes the claim that $2^p - 1$ is composite for the other 44 primes p which do not exceed 257. How Mersenne reached his conclusions we shall never know, but it is clear that with what was available at that time he could not have tested primality of each of the numbers by hand calculation. In fact it took 303 years and a lot of labour to check out the whole of Mersenne's claim. It is remarkable how close to the truth his claim was, for we now know that he made just five mistakes.

In 1732 Euler confirmed that M_{29} is composite and 40 years later he established that M_{31} is indeed prime by eliminating all possible divisors. For some 200 years M_{19} had stood as the largest known prime number but now a much larger one, M_{31}, was to hand.

To keep track of how the even perfect numbers are coming along, M_{19} and M_{31} provide us with the seventh and eighth perfect numbers:

$$2^{18}(2^{19} - 1) = 137\,438\,691\,328;$$
$$2^{30}(2^{31} - 1) = 2\,305\,843\,008\,139\,952\,128.$$

By 1811 no larger prime number had been found and a distinguished mathematician of that time, Barlow, asserted pessimistically in his book *Theory of Numbers*, that 'M_{31} is the greatest prime that will ever be found'. His reasoning was that the search for larger primes had curiosity value rather than real use, and the sheer size of M_{31} would deter a search for anything larger. How wrong he was. A mere 65 years later the French mathematician Edouard Lucas proved that

$$M_{127} = 170\,141\,183\,460\,469\,231\,731\,687\,303\,715\,884\,105\,727$$

is prime, and with it a ninth perfect number, $2^{126}(2^{127} - 1)$, which has 77 digits, had been found. Lucas also uncovered the first mistake in Mersenne's list when he proved that M_{67} is in fact composite. His proof demonstrated the lack of primality without producing the actual divisors. That had to wait until 1903 when, at a meeting of the American Mathematical Society a professor F.N. Cole delivered a talk entitled 'On the factorization of large numbers'. His 'talk' consisted of calculating $2^{67} - 1$ by hand on one half of the blackboard and then, on the other half, working out the product

$$193\,707\,721 \times 761\,838\,257\,287.$$

The two agreed. He sat down to prolonged applause having said not a single word!

By 1914 three omissions in Mersenne's list had been found. M_{61}, M_{89} and M_{107} had all been shown to be prime. But M_{127} remained the largest known Mersenne prime until 1952 when an American, R. Robinson, using what was then regarded as a large computer, showed that M_{521}, M_{607}, M_{1279}, M_{2203} and M_{2281} are all prime. As more advanced computers have subsequently been brought to bear on the problem, further Mersenne primes have been discovered. As we approach the mid-1990's, the list has reached 32 Mersenne primes, and therefore 32 perfect numbers. These are given in the following table.

Though M_{127} was the ninth Mersenne prime discovered, it is the twelfth in ascending order.

Table 2.2 Mersenne primes

Prime p	Number of digits in M_p	Perfect number	Date discovered
2	1	$2(2^2 - 1)$	Known to Euclid
3	1	$2^2(2^3 - 1)$	Known to Euclid
5	2	$2^4(2^5 - 1)$	Known to Euclid
7	3	$2^6(2^7 - 1)$	Known to Euclid
13	4	$2^{12}(2^{13} - 1)$	1456
17	6	$2^{16}(2^{17} - 1)$	1588
19	6	$2^{18}(2^{19} - 1)$	1588
31	10	$2^{30}(2^{31} - 1)$	1772
61	19	$2^{60}(2^{61} - 1)$	1883
89	27	$2^{88}(2^{89} - 1)$	1911
107	33	$2^{106}(2^{107} - 1)$	1914
127	39	$2^{126}(2^{127} - 1)$	1876
521	157	$2^{520}(2^{521} - 1)$	1952
607	183	$2^{606}(2^{607} - 1)$	1952
1279	386	$2^{1278}(2^{1279} - 1)$	1952
2203	664	$2^{2202}(2^{2203} - 1)$	1952
2281	687	$2^{2280}(2^{2281} - 1)$	1952
3217	969	$2^{3216}(2^{3217} - 1)$	1957
4253	1281	$2^{4252}(2^{4253} - 1)$	1961
4423	1332	$2^{4422}(2^{4423} - 1)$	1961
9689	2917	$2^{9688}(2^{9689} - 1)$	1963
9941	2993	$2^{9940}(2^{9941} - 1)$	1963
11 213	3376	$2^{11\,212}(2^{11\,213} - 1)$	1963
19 937	6002	$2^{19\,936}(2^{19\,937} - 1)$	1971
21 701	6533	$2^{21\,700}(2^{21\,701} - 1)$	1978
23 209	6987	$2^{23\,208}(2^{23\,209} - 1)$	1979
44 497	13 395	$2^{44\,496}(2^{44\,497} - 1)$	1979
86 243	25 962	$2^{86\,242}(2^{86\,243} - 1)$	1982
110 503	33 265	$2^{110\,502}(2^{110\,503} - 1)$	1990
132 049	39 751	$2^{132\,048}(2^{132\,049} - 1)$	1983
216 091	65 050	$2^{216\,090}(2^{216\,091} - 1)$	1985
756 839	227 832	$2^{756\,838}(2^{756\,839} - 1)$	1993

2.3 Some results

From the unfolding story you might not be surprised to learn that we do not know yet whether or not there exist infinitely many Mersenne primes, and hence infinitely many even perfect numbers. Curiously we are no further forward in the opposite direction. It remains unknown whether or not infinitely many of the Mersenne numbers M_p are composite. One hope for the infinitude of the Mersenne primes arose from the observation that for the first four such numbers, namely 3, 7, 31 and 127, the numbers M_3, M_7, M_{31}, and M_{127} had also turned out to be prime. Could it be generally true that if M_p is prime then M_{M_p} is prime? If so we would immediately have infinitely many Mersenne primes. But this hope was dashed when, in 1953, a fast computer confirmed that

$$M_{M_{13}} = 2^{M_{13}} - 1 = 2^{8191} - 1$$

is composite.

As the challenge to determine the primality or otherwise of M_p has persisted, a variety of methods and results for attacking particular values of p have emerged. As illustrations we shall introduce a couple of the simpler ones here.

Problem 2.2 _____

Suppose that p and $q = 2p + 1$ are both prime. By applying FLT to 2^{q-1} show that either q divides M_p or q divides $M_p + 2$.

_____

The result of Problem 2.2 can be followed through to the conclusion that q divides M_p when $p \equiv 3 \pmod 4$. We shall record this result but postpone the proof of the missing final step until *Unit 6*, when it can be deduced quickly from results established there.

Theorem 2.2

If p and $2p + 1$ are primes, where $p \equiv 3 \pmod 4$, then $2p + 1$ divides M_p.

To illustrate the use of Theorem 2.2, note that $11 \equiv 3 \pmod 4$ and $2 \times 11 + 1 = 23$ is prime. We conclude that 23 is a divisor of M_{11}, a fact which we already knew. But in just the same way, Theorem 2.2, together with a brief glance at a table of primes, tells us that 47 is a divisor of M_{23}, 167 is a divisor of M_{83}, 263 is a divisor of M_{131}, 359 is a divisor of M_{179}, 479 is a divisor of M_{239} and 503 is a divisor of M_{251}. So this one result has confirmed seven of the composite values in Mersenne's list.

The second result, which was known to Fermat, is not quite so dramatic but it severely restricts the potential divisors of a given M_p.

Theorem 2.3

Any prime divisor of M_p, where p is an odd prime, is of the form $2kp + 1$ for some positive integer k.

Proof of Theorem 2.3

Let q be a prime divisor of $M_p = 2^p - 1$. Then $2^p \equiv 1 \pmod q$. But we also have, from FLT, that $2^{q-1} \equiv 1 \pmod q$. Now let the order of 2 modulo q be c, so that both p and $q - 1$ are multiples of c. But as p is a prime and $c > 1$ (since $2^1 \equiv 1 \pmod q$ is not possible), it follows that $c = p$. Hence $q - 1$ is a multiple of p.

See Theorem 2.3 of *Unit 4*.

Finally, writing $q - 1 = tp$, for some integer t, gives $q = tp + 1$. But if the integer t is odd then the prime q would be even, which certainly cannot be the case. Hence t is a positive even integer, and putting $t = 2k$ gives $q = 2kp + 1$, as claimed. ∎

To illustrate how Theorem 2.3 might be applied, consider M_{19}. The prime divisors of M_{19} must be of the form $38k + 1$, and if M_{19} is not itself prime then it must have a divisor which does not exceed the square-root of M_{19} (this is approximately $2^{19/2}$ which is close to 724). There are nineteen numbers of the form $38k + 1$, with k a positive integer, which do not exceed 724, namely $39, 77, 115, 153, \ldots, 723$, and reference to a table of primes reveals that just six of these are prime, namely 191, 229, 419, 457, 571 and 647.

Hence, to establish the primality of M_{19}, it is sufficient to show that it does not have one of these six primes as a divisor.

Applying Theorem 2.3 to test the primality of M_{17}, it can be shown that, to be composite, M_{17} must be divisible by one of four primes. Find these four primes.

_____

We have concentrated on the even perfect numbers with good reason. No odd perfect number is known; it is generally believed that odd numbers cannot be perfect, but nobody has managed to prove this. The search for odd perfect numbers has, however, not been without interest. Euler made the first real start when he proved the following result concerning the prime decomposition that an odd perfect number must have.

Theorem 2.4 Euler's form for an odd perfect number

If n is an odd perfect number then

$$n = p^k p_1^{2k_1} p_2^{2k_2} \ldots p_r^{2k_r},$$

where the p_is are distinct primes and $p \equiv k \equiv 1 \pmod 4$.

The exceptional prime p has been listed first but is not necessarily the smallest prime in the decomposition.

Proof of Theorem 2.4 is beyond the scope of this course.

This result says that with one exception all the exponents in the prime decomposition of n must be even, and the exceptional prime and its exponent are both of the form $4t + 1$ (though not necessarily for the same t). This result has, nowadays, been developed much further. For example, it is now known that the total number of distinct primes involved must be at least eight, one of the prime divisors must be greater than $100\,000$ and the number n itself must have more than 300 digits.

We cannot leave this subject matter without saying a few words about the *Fermat numbers*. Along with the Mersenne numbers, which have the form $2^n - 1$, it seems natural to consider numbers of the form $2^n + 1$. Can such numbers be prime? Fermat was aware that if the exponent n had an odd divisor (other than 1) then $2^n + 1$ must be composite, as can be seen from the identity

$$2^{(2r+1)s} + 1 = (2^s + 1)\left(2^{2rs} - 2^{(2r-1)s} + 2^{(2r-2)s} - \cdots - 2^s + 1\right).$$

To have a chance of being prime the exponent n must therefore be a power of 2, and so Fermat turned his attention to numbers of the form $2^{2^n} + 1$.

Definition 2.3 Fermat numbers

The numbers $F_n = 2^{2^n} + 1$, where n is a non-negative integer, are known as *Fermat numbers*.

2^{2^n} means $2^{(2^n)}$, not $\left(2^2\right)^n$.

The first five Fermat numbers,

$$F_0 = 3, \quad F_1 = 5, \quad F_2 = 17, \quad F_3 = 257 \quad \text{and} \quad F_4 = 65\,537,$$

are primes.

In a letter to Mersenne, Fermat expressed his belief that all the numbers F_n are prime, although he admitted that he could not provide a proof. It was Euler who provided the counter-example: that 641 divides F_5. Since then all the numbers F_n, for n from 5 to 21, have been shown to be composite. The prime decomposition:

$$F_6 = 274\,177 \times 67\,280\,421\,310\,721$$

was known in 1880, but it required electronic computers to uncover, in 1970, the next one:

$$F_7 = 59\,649\,589\,127\,497\,217 \times 5\,704\,689\,200\,685\,129\,054\,721.$$

The prime decomposition of F_8 was found in 1980, followed by that of F_{11} in 1989. F_{11} has two relatively small prime divisors:

$$F_{11} = 319\,489 \times 974\,849 \times p_1 \times p_2 \times p_3,$$

where the primes p_1, p_2 and p_3 have 21, 22 and 564 digits respectively. The prime decomposition of F_9 was found in 1990. The method used to achieve this involved breaking the problem into many smaller division problems which were tackled simultaneously by hundreds of computers around the world. Nevertheless it took two months to complete the task!

The prime decomposition of F_{10} is still not known, nor indeed is it for any F_n, $n \geq 12$. However, at least one prime divisor of F_n is known for $n = 10$, 12, 13, 15, 16, 17, 18, 19, 21, 23 and some larger values. Of the values missing here, F_{14} and F_{20} are known to be composite although no actual prime divisor has been found for either. That leaves F_{22} as the next Fermat number which may yet turn out to be prime.

Many more Fermat numbers have been shown to be composite by various methods which often hinge on exhibiting some prime which must be a divisor. The largest Fermat number known to be composite at the present time is F_{23471} which was shown, in 1984, to be divisible by $5 \times 2^{23473} + 1$. Compared with the size of F_{23471} itself, this is a very, very small divisor, and yet it is a number with over 7000 digits!

3 EULER'S ϕ-FUNCTION

3.1 Euler's Theorem

Let us look again at the *order* of an integer. In Section 2 of *Unit 4* we had the following definition.

> If p is prime and $\gcd(a, p) = 1$ then the order of a modulo p is the least positive integer c such that $a^c \equiv 1 \pmod{p}$.

In this section we are going to investigate the notion of *order* for composite moduli. In the table below we have calculated successive powers of integers a, modulo 10, in an attempt to find the order of a by discovering a power which is congruent to 1.

Table 3.1 Powers modulo 10

a	1	2	3	4	5	6	7	8	9
a^2	1	4	9	6	5	6	9	4	1
a^3	1	8	7	4	5	6	3	2	9
a^4	1	6	1	6	5	6	1	6	1
a^5	1	2	3	4	5	6	7	8	9

We have stopped at the fifth power in the realization that the value 1 is not going to arise for $a = 2, 4, 5, 6$ and 8, where the powers are already cycling around values other than 1. What the table tells us is that the numbers 1, 3, 7 and 9 have respective orders 1, 4, 4 and 2 modulo 10, but the concept of order does not apply to the remaining integers. We note that the four which have an order are precisely the four which are relatively prime to the modulus 10.

Let us turn now to a general modulus n and suppose that the number a has an order modulo n, that is, there is an integer $r \geq 1$ such that $a^r \equiv 1 \pmod{n}$. This means that $a^r - 1 = kn$, for some integer k. Now $\gcd(a, n)$ divides a, and therefore a^r, and it divides n, and so divides $a^r - kn = 1$. Thus $\gcd(a, n) = 1$, and so a is relatively prime to n.

All this suggests that we should be concentrating on the set of least positive residues of n which are relatively prime to n. Before defining this set it will be useful to define the following number-theoretic function which gives the size of this set.

Definition 3.1 Euler's ϕ-function

For each integer $n \geq 1$ define $\phi(n)$ to be the number of positive integers not exceeding n which are relatively prime to n.

Note that $\phi(1) = 1$ since $\gcd(1, 1) = 1$.

For example, the integers a in the range $1 \leq a \leq 10$ for which $\gcd(a, 10) = 1$ are 1, 3, 7 and 9, four in number, and so $\phi(10) = 4$. In terms of residue classes, $\phi(n)$ is the number of classes which consist of integers relatively prime to n. Selection of one member from each such class produces a *reduced set of residues*.

Definition 3.2

A *reduced set of residues modulo n* is a set of integers $\{a_1, a_2, a_3, \ldots, a_{\phi(n)}\}$ each of which is relatively prime to n and no two of which are congruent modulo n.

The reduced set of residues in which each is a positive integer less than or equal to n is called the *reduced set of least positive residues modulo n*.

As $\gcd(0, n) = n$, 0 does not appear in a reduced set of residues modulo n for any $n > 1$.

Problem 3.1 _____

Complete the following table giving, for $n = 1$ to 10, the reduced set of least positive residues modulo n and the value of $\phi(n)$.

n	Reduced set of residues	$\phi(n)$
1		
2		
3		
4		
5		
6		
7		
8		
9		
10	$\{1, 3, 7, 9\}$	4

To determine $\phi(n)$, for $n > 1$, we need to know how many numbers in the set $\{1, 2, 3, \ldots, n-1\}$ are relatively prime to n. If n happens to be prime they are all relatively prime to n, and so $\phi(n) = n - 1$. On the other hand, if n is composite then one of the listed numbers is a divisor $d > 1$ of n, and as $\gcd(d, n) = d$, it is not relatively prime to n. In this case $\phi(n) \leq n - 2$. The conclusion from these remarks is that n is prime if, and only if, $\phi(n) = n - 1$.

Since $n > 1$ we may exclude n from consideration since
$$\gcd(n, n) = n \neq 1.$$

We shall return to investigate properties of the ϕ-function shortly, but in the meantime let us head straight for our main result of this section: Euler's generalization of FLT. For a composite modulus n, and with attention restricted to integers a relatively prime to n, we are interested in finding exponents c such that $a^c \equiv 1 \pmod{n}$. The smallest positive integer c satisfying this congruence will be the order of a modulo n. To garner more numerical evidence we have calculated various powers of a reduced set of residues modulo 9 and the results are shown in the table below.

Table 3.2 Powers modulo 9

a	1	2	4	5	7	8
a^2	1	4	7	7	4	1
a^3	1	8	1	8	1	8
a^4	1	7	4	4	7	1
a^5	1	5	7	2	4	8
a^6	1	1	1	1	1	1

What should first catch the eye about this table is the row of sixth powers: $a^6 \equiv 1 \pmod 9$ for each integer a relatively prime to 9. A shrewed observer will also have noted that the exponent 6 just happens to be $\phi(9)$. There is nothing special about the number 9; what is illustrated here is generally true.

Theorem 3.1 Euler's Theorem—a generalization of FLT

If n is a positive integer and a is any integer with $\gcd(a, n) = 1$, then $a^{\phi(n)} \equiv 1 \pmod n$.

Notice that when n is a prime p, then, since $\phi(p) = p - 1$, this result says that $a^{p-1} \equiv 1 \pmod p$, which is precisely FLT. So this result is indeed a generalization.

The proof that follows ought to look familiar, for it is little more that a re-run of our proof of FLT. However, where the latter worked with a set of non-zero residues modulo p, we now base the construction on a reduced set of residues modulo n.

Proof of Theorem 3.1

Let $\{b_1, b_2, b_3, \ldots, b_{\phi(n)}\}$ be a reduced set of residues modulo n. Consider the set obtained by multiplying each by a:

$$\{ab_1, ab_2, ab_3, \ldots, ab_{\phi(n)}\}.$$

We claim that this is also a reduced set of residues modulo n. Certainly no two of the $\phi(n)$ integers in this set are congruent modulo n because from $ab_i \equiv ab_j \pmod n$ and $\gcd(a, n) = 1$, cancellation would give $b_i \equiv b_j \pmod n$, which can only occur when $i = j$. Moreover each integer of the set is relatively prime to n because, if some prime p divides both ab_i and n then, appealing to Euclid's Lemma, p divides either a or b_i. But p cannot divide both a and n for that would contradict $\gcd(a, n) = 1$, and similarly p cannot divide both b_i and n because that contradicts $\gcd(b_i, n) = 1$. It follows that such a prime p cannot exist and therefore $\gcd(ab_i, n) = 1$.

As each of the two sets is a reduced set of residues modulo n their products are congruent modulo n, that is,

Each b_i is congruent modulo n to some ab_j.

$$ab_1 ab_2 ab_3 \ldots ab_{\phi(n)} \equiv b_1 b_2 b_3 \ldots b_{\phi(n)} \pmod n.$$

Now since $\gcd(b_i, n) = 1$, we can cancel each b_i in turn to obtain $a^{\phi(n)} \equiv 1 \pmod n$, as claimed. ∎

To illustrate the above proof let us trace it through for the case $n = 15$ and $a = 7$. We obtain a reduced set of residues modulo 15 by taking the set of least positive residues of 15 and eliminating those which are multiples of either 3 or 5, leaving the set of $\phi(15) = 8$ integers, $\{1, 2, 4, 7, 8, 11, 13, 14\}$. The proof showed that we obtain the same set of eight integers, modulo 15, when we multiply each by 7. In fact

$$7 \times 1 \equiv 7, \quad 7 \times 2 \equiv 14, \quad 7 \times 4 \equiv 13, \quad 7 \times 7 \equiv 4,$$
$$7 \times 8 \equiv 11, \quad 7 \times 11 \equiv 2, \quad 7 \times 13 \equiv 1, \quad 7 \times 14 \equiv 8.$$

Therefore

$$7^8 \times (1 \times 2 \times 4 \times 7 \times 8 \times 11 \times 13 \times 14) \equiv$$
$$1 \times 2 \times 4 \times 7 \times 8 \times 11 \times 13 \times 14 \ (\text{mod } 15),$$

and after cancellation by the least positive residues, which is permissible since each of the eight numbers is relatively prime to 15, we obtain $7^8 \equiv 1 \ (\text{mod } 15)$.

Problem 3.2

Determine $\phi(14)$ and confirm that

$$3^{\phi(14)} \equiv 1 \ (\text{mod } 14) \quad \text{and} \quad 5^{\phi(14)} \equiv 1 \ (\text{mod } 14).$$

3.2 The order of an integer to a composite modulus

With the benefit of Euler's Theorem we are now in a position to give a general definition of the *order* of a number.

> ### Definition 3.3 The order of an integer modulo n
>
> If $n \geq 1$ and $\gcd(a, n) = 1$, then the *order* of a modulo n is the least positive integer c such that $a^c \equiv 1 \ (\text{mod } n)$.

We were already aware that the condition $\gcd(a, n) = 1$ is necessary for the existence of an integer c with $a^c \equiv 1 \ (\text{mod } n)$. Euler's Theorem shows that the condition is also sufficient, since $a^{\phi(n)} \equiv 1 \ (\text{mod } n)$. So $c \leq \phi(n)$.

Since $a^k \equiv b^k \ (\text{mod } n)$ whenever $a \equiv b \ (\text{mod } n)$, it can be seen that congruent integers have the same order. Hence to find all orders modulo n it suffices to find the orders of the $\phi(n)$ integers in a reduced set of residues modulo n.

Problem 3.3

Find all the possible orders of integers modulo 14.

In *Unit 4*, we established that if $a^k \equiv 1 \ (\text{mod } p)$, for some prime p, then k must be a multiple of the order of a. It should come as no surprise to learn that this result also generalizes.

Theorem 2.3 of *Unit 4*.

> ### Theorem 3.2
>
> If the integer a has order c modulo n then $a^k \equiv 1 \ (\text{mod } n)$ if, and only if, k is a multiple of c. In particular the order c divides $\phi(n)$.

Proof of Theorem 3.2

If k is a multiple of c, say $k = cr$, then

$$a^k = a^{cr} = (a^c)^r \equiv 1^r \equiv 1 \;(\text{mod } n).$$

On the other hand, suppose that $a^k \equiv 1 \;(\text{mod } n)$. Dividing k by c, the Division Algorithm gives integers q and r such that $k = qc + r$, where $0 \le r < c$. Then

$$1 \equiv a^k \equiv a^{qc+r} \equiv (a^c)^q a^r \equiv (1)^q a^r \equiv a^r \;(\text{mod } n).$$

This forces $r = 0$, for otherwise $a^r \equiv 1 \;(\text{mod } n)$ contradicts c being the least positive integer with the property $a^c \equiv 1 \;(\text{mod } n)$. So $k = qc$, and k is a multiple of c, as claimed.

As $a^{\phi(n)} \equiv 1 \;(\text{mod } n)$, it follows that $\phi(n)$ is a multiple of the order c. ∎

Theorem 3.2 simplifies the computational task of determining the order of an integer modulo n because it tells us that we can confine attention to those exponents which are divisors of $\phi(n)$.

Example 3.1

Determine the order of 5 modulo 38.

To find $\phi(38)$ it suffices to count how many integers in the set $\{1, 2, 3, \ldots, 36, 37\}$ are relatively prime to 38. As 38 has prime decomposition 2×19 the integers relatively prime to 38 are precisely those which are not multiples of either 2 or 19. This leaves the odd numbers with the exception of 19, of which there are eighteen. So $\phi(38) = 18$. The order of 5 modulo 38 is therefore a divisor of 18 and consequently is one of 1, 2, 3, 6, 9 and 18. Working modulo 38 we have

$$5^1 = 5;$$
$$5^2 = 25;$$
$$5^3 = 125 \equiv 11;$$
$$5^6 = 5^3 \times 5^3 \equiv 11 \times 11 \equiv 7;$$
$$5^9 = 5^6 \times 5^3 \equiv 7 \times 11 \equiv 1.$$

Hence the order of 5 modulo 38 is 9. ◆

Problem 3.4

Determine:

(a) the order of 4 modulo 21;

(b) the order of 7 modulo 22.

3.3 Applications of Euler's Theorem

We have seen numerous applications of FLT which assist in the determination of remainders when large exponents are involved. Euler's Theorem can be used in the same way, but with the extra dimension that we need not be working to a prime modulus.

Example 3.2

Determine the remainder when 23^{23} is divided by 15.

Since $\gcd(23, 15) = 1$ and $\phi(15) = 8$, Euler's Theorem tells us that
$23^8 \equiv 1 \pmod{15}$.

Therefore

$$23^{23} = (23^8)^2 \times 23^7 \equiv 23^7 \equiv 8^7 \equiv 64^3 \times 8$$
$$\equiv 4^3 \times 8 \equiv 16 \times 4 \times 8 \equiv 1 \times 32 \equiv 2 \pmod{15},$$

and the remainder on dividing 23^{23} by 15 is 2.   ◆

Problem 3.5

Confirm that $\phi(100) = 40$ and hence determine the final two digits of the
following numbers.

(a) 17^{83}

(b) 19^{39} *Hint:* Perhaps consider the congruence $19x \equiv 1 \pmod{100}$.

The sequence of integers

$$1, \ 11, \ 111, \ 1\,111, \ 11\,111, \ldots,$$

the nth term of which is the number comprising n 1's, has aroused a great
deal of interest. The numbers in this sequence are called *repunits* and, just
as with the Mersenne numbers, much effort has been expended on attempts
to factorize them. If we denote the repunit with n 1's by R_n then it turns out
that if n is composite so too is R_n. In fact if $n = rs$ then each of R_r and R_s
is a divisor of R_{rs}, as can be seen from the result of Problem 3.6 below.

For example,

$$R_6 = 111\,111$$
$$= 11 \times 10\,101 = R_2 \times 10\,101$$
$$= 111 \times 1001 = R_3 \times 1001.$$

Prime repunits are rare. Table 3.3 shows the prime decompositions of R_p,
for primes p up to 30, and you will notice just three of these turn out to be
prime.

Table 3.3   Prime decompositions of small repunits

p	R_p
2	11
3	3×37
5	41×271
7	239×4649
11	$21\,649 \times 513\,239$
13	$53 \times 79 \times 265\,371\,653$
17	$2\,071\,723 \times 5\,363\,222\,357$
19	$1\,111\,111\,111\,111\,111\,111$
23	$11\,111\,111\,111\,111\,111\,111\,111$
29	$3191 \times 16\,763 \times 43\,037 \times 62\,003 \times 77\,843\,839\,397$

Before progressing let us record two simple facts about repunits which will
prove useful in what follows. First, suppose we multiply the repunit R_n by 9:

$$9R_n = \overbrace{999\ldots99}^{n \ 9\text{'s}} = 10^n - 1.$$

This gives us a new expression for R_n shown below.

Repunit property 1

$$R_n = \frac{10^n - 1}{9}$$

Second, consider the repunit R_{m+n}.

$$R_{m+n} = \overbrace{111\ldots 1}^{m\text{ 1's}}\overbrace{1\ldots 11}^{n\text{ 1's}}$$

$$= \overbrace{111\ldots 1}^{m\text{ 1's}}\overbrace{0\ldots 00}^{n\text{ 0's}} + \overbrace{1\ldots 11}^{n\text{ 1's}}$$

$$= R_m \times 10^n + R_n.$$

Repunit property 2

$$R_{m+n} = R_m \times 10^n + R_n$$

If you were not convinced by the argument leading to property 2 you can prove it in a straightforward fashion by substituting the expressions for R_{m+n}, R_m and R_n given by property 1.

One curiosity associated with repunits can be explained via Euler's Theorem. Every odd integer which is not divisible by 5 has a multiple which is a repunit.

Example 3.3

Show that if the integer n is relatively prime to 10 then n has a multiple each of whose digits is a 1.

Suppose that n is such that $\gcd(n, 10) = 1$. Since $\gcd(9, 10) = 1$ we have that $\gcd(9n, 10) = 1$ and so we can apply Euler's Theorem:

$$10^{\phi(9n)} \equiv 1 \pmod{9n},$$

that is,

$$10^{\phi(9n)} - 1 = k \times 9n, \quad \text{for some integer } k.$$

Dividing both sides of this equation by 9 we obtain

$$R_{\phi(9n)} = kn,$$

which confirms the claim that n has a multiple which is a repunit, namely $R_{\phi(9n)}$. ◆

Problem 3.6

Use mathematical induction to prove that R_{mn} is divisible by R_n, for all integers $m \geq 1$.

Problem 3.7

Let $p > 5$ be prime.

(a) Show that p divides R_{p-1}. *Hint*: Consider the number $10^{p-1} - 1$.

(b) Use the Division Algorithm to show that if R_n is the smallest repunit which is a multiple of p, then n divides $p - 1$.

(c) Find the smallest repunit which is a multiple of (i) 7 and (ii) 13.

We shall go through one final example using Euler's Theorem.

Example 3.4

Show that if $\gcd(n, 42) = 1$ then $n^6 \equiv 1 \pmod{7 \times 8 \times 9}$.

Using what should be a familiar approach, we aim to show that n^6 is congruent to 1 for each of the pairwise relatively prime moduli 7, 8 and 9. Since $\phi(7) = 6$ and $\phi(9) = 6$, Euler's Theorem gives

$n^6 \equiv 1 \pmod 7$, whenever $\gcd(n, 7) = 1$, and

$n^6 \equiv 1 \pmod 9$, whenever $\gcd(n, 9) = 1$, that is, whenever $\gcd(n, 3) = 1$.

Finally, since the square of any odd integer is congruent modulo 8 to 1 we have

$n^6 \equiv 1 \pmod 8$, for all integers n such that $\gcd(n, 2) = 1$.

Now the given condition $\gcd(n, 42) = 1$ amounts to $\gcd(n, 2) = 1$, $\gcd(n, 3) = 1$ and $\gcd(n, 7) = 1$, so all three congruences hold and $n^6 \equiv 1 \pmod{7 \times 8 \times 9}$ follows.   ♦

Any odd integer is of the form $2n + 1$ and
$$(2n + 1)^2 = 4n^2 + 4n + 1$$
$$= 4n(n + 1) + 1.$$
As either n or $n + 1$ is even, the square is congruent modulo 8 to 1.

Problem 3.8 _____

Show that $n^{33} \equiv n \pmod{15 \times 16 \times 17}$ for every odd integer n.

---

4 PROPERTIES OF EULER'S ϕ-FUNCTION

4.1 The multiplicativity of Euler's ϕ-function

We have evaluated $\phi(n)$ for several small values of n and we have discovered the one general result that $\phi(n) = n - 1$ if, and only if, n is prime. Any hope of finding a general formula for $\phi(n)$ in terms of the prime decomposition of n, such as we found for $\sigma(n)$ and $\tau(n)$, would be heightened if it happens that ϕ is a multiplicative function. Fortunately it is.

> **Theorem 4.1 Multiplicativity of ϕ**
>
> The function ϕ is multiplicative.

The underlying idea in the proof that we are about to give, showing that $\phi(mn) = \phi(m)\phi(n)$ for relatively prime integers m and n, is very simple. From a reduced set of residues modulo m and a reduced set of residues modulo n we shall construct a set of $\phi(m)\phi(n)$ numbers which will be a reduced set of residues modulo mn. To illustrate this, consider the case $m = 6$ and $n = 7$. Reduced sets of residues for these moduli are $A = \{1, 5\}$ for modulus 6 and $B = \{1, 2, 3, 4, 5, 6\}$ for modulus 7. Now consider the set of $\phi(6)\phi(7) = 12$ integers

$\{na + mb : a \in A, b \in B\} = \{7a + 6b : a \in A, b \in B\}$.

This set is

$\{13, 19, 25, 31, 37, 43, 41, 47, 53, 59, 65, 71\}$,

which, on reducing each number modulo 6×7 and putting the terms into increasing order, is

$\{1, 5, 11, 13, 17, 19, 23, 25, 29, 31, 37, 41\}$.

The first six listed values correspond to $a = 1$ and the second six to $a = 5$.

Each of these integers is relatively prime to 42, and indeed this is a reduced set of residues for modulus 42, showing that $\phi(6 \times 7) = \phi(6)\phi(7)$.

The proof which follows reproduces this construction for general relatively prime moduli m and n.

Proof of Theorem 4.1

Let m and n be relatively prime integers; our task is to show $\phi(mn) = \phi(m)\phi(n)$.

Let $\{r_1, r_2, r_3, \ldots, r_{\phi(m)}\}$ and $\{s_1, s_2, s_3, \ldots, s_{\phi(n)}\}$ be reduced sets of residues for the respective moduli m and n. We show that the set of $\phi(m)\phi(n)$ integers

$$\{nr_i + ms_j : 1 \leq i \leq \phi(m), 1 \leq j \leq \phi(n)\}$$

is a reduced set of residues for modulus mn, thereby showing that $\phi(m)\phi(n) = \phi(mn)$. There are three properties to be established:

(a) each integer in this set is relatively prime to mn.

(b) no two integers in this set are congruent modulo mn;

(c) each integer relatively prime to mn is congruent modulo mn to an integer in this set;

We prove each of these in turn.

(a) Suppose that p is a prime divisor of $\gcd(nr_i + ms_j, mn)$. From p divides mn with $\gcd(m, n) = 1$ we conclude that either p divides m or p divides n but not both. Let us suppose that p divides m. Then, as p divides $nr_i + ms_j$ we must have p divides nr_i and therefore p divides r_i. But then p divides $\gcd(m, r_i) = 1$, a contradiction. The alternative supposition that p divides n leads to the similar contradiction that p divides $\gcd(n, s_j)$. No such prime p can exist and so each of the integers $nr_i + ms_j$ is relatively prime to mn.

> Remember that r_i belongs to a reduced set of residues modulo m.

(b) Suppose that $nr + ms \equiv nr' + ms' \pmod{mn}$, where r, r', s and s' are chosen from the reduced sets of residues. Then

$$n(r - r') + m(s - s') = k(mn), \text{ for some integer } k.$$

As m divides two of the terms in this equation it must divide the third, that is, m divides $n(r - r')$. Now $\gcd(m, n) = 1$, and so Euclid's Lemma tells that m divides $r - r'$, or what amounts to the same thing, $r \equiv r' \pmod{m}$. But r and r' are members of a reduced set of residues modulo m, and so $r = r'$. The same argument shows that $s = s'$ and we conclude that distinct elements of the given set cannot be congruent modulo mn.

(c) Let k be any integer relatively prime to mn. Since $\gcd(m, n) = 1$, we can write $k = nr + ms$, for some integers r and s. Now suppose there is a prime p which divides both m and r; such a prime would be a common divisor of k and mn, contradicting $\gcd(k, mn) = 1$. Hence r is relatively prime to m and as such is congruent modulo m to some r_i, for $1 \leq i \leq \phi(m)$. And, by exactly the same argument with the roles of m and n interchanged, s is congruent modulo n to one of the s_j in the reduced set. Writing $r = r_i + am$ and $s = s_j + bn$ we have

> See Theorem 4.3 of *Unit 1*.

$$k = nr + ms = nr_i + ms_j + mn(a + b) \equiv nr_i + ms_j \pmod{mn}$$

as required.

As (a), (b) and (c) are all established the proof is complete. ∎

Problem 4.1

Use the multiplicative property of ϕ to evaluate $\phi(24)$, $\phi(70)$ and $\phi(420)$.

4.2 The formula for $\phi(n)$

To obtain a formula which gives the value of $\phi(n)$ in terms of the prime decomposition of n all we need now is to be able to evaluate ϕ at a general prime power and appeal to the multiplicative property.

Theorem 4.2 Formula for $\phi(n)$

If n has prime decomposition $n = p_1^{k_1} p_2^{k_2} \ldots p_r^{k_r}$ then,

$$\phi(n) = p_1^{k_1} p_2^{k_2} \ldots p_r^{k_r} \left(1 - \frac{1}{p_1}\right) \left(1 - \frac{1}{p_2}\right) \ldots \left(1 - \frac{1}{p_r}\right)$$

$$= n \prod_{p|n} \left(1 - \frac{1}{p}\right).$$

The formula does not apply for $n = 1$, where $\phi(1) = 1$.

There is one term in the product $\prod$ for each prime which divides n, that is, for $p_1, p_2, \ldots, p_r$.

Proof of Theorem 4.2

To determine $\phi(p^k)$ we count the number of integers lying between 1 and p^k inclusive which are relatively prime to p^k. As the only divisors of p^k are powers of p, $\gcd(n, p^k) = 1$ if, and only if, p does not divide n. There are p^{k-1} multiples of p lying between 1 and p^k, namely

$$p, 2p, 3p, \ldots, p^{k-1}p,$$

and the remaining $p^k - p^{k-1}$ integers are relatively prime to p. Hence,

$$\phi(p^k) = p^k - p^{k-1} = p^k \left(1 - \frac{1}{p}\right).$$

The multiplicative property of ϕ now gives

$$\begin{aligned}
\phi(n) &= \phi\left(p_1^{k_1} p_2^{k_2} \ldots p_r^{k_r}\right) \\
&= \phi\left(p_1^{k_1}\right) \phi\left(p_2^{k_2}\right) \ldots \phi\left(p_r^{k_r}\right) \\
&= p_1^{k_1} \left(1 - \frac{1}{p_1}\right) p_2^{k_2} \left(1 - \frac{1}{p_2}\right) \ldots p_r^{k_r} \left(1 - \frac{1}{p_r}\right) \\
&= p_1^{k_1} p_2^{k_2} \ldots p_r^{k_r} \left(1 - \frac{1}{p_1}\right) \left(1 - \frac{1}{p_2}\right) \ldots \left(1 - \frac{1}{p_r}\right) \\
&= n \prod_{p|n} \left(1 - \frac{1}{p}\right). \qquad \blacksquare
\end{aligned}$$

With the aid of this formula computing values of $\phi(n)$ presents little problem. For example,

$$\begin{aligned}
\phi(7920) &= \phi\left(2^4 \times 3^2 \times 5 \times 11\right) \\
&= 2^4 \times 3^2 \times 5 \times 11 \times \left(1 - \frac{1}{2}\right)\left(1 - \frac{1}{3}\right)\left(1 - \frac{1}{5}\right)\left(1 - \frac{1}{11}\right) \\
&= 2^4 \times 3^2 \times 5 \times 11 \times \frac{1}{2} \times \frac{2}{3} \times \frac{4}{5} \times \frac{10}{11} \\
&= 2^7 \times 3 \times 5 \\
&= 1920.
\end{aligned}$$

Problem 4.2 _____

Determine $\phi(336)$ and $\phi(8p^3)$, where p is any odd prime.

The two examples which follow make use of the multiplicative nature of ϕ and employ the formula above.

Example 4.1

If p is a prime and $n \geq 1$ prove that

$$\phi(pn) = \begin{cases} (p-1)\phi(n), & \text{if } p \text{ does not divide } n, \\ p\phi(n), & \text{if } p \text{ divides } n. \end{cases}$$

Deduce that if d divides n then $\phi(d)$ divides $\phi(n)$.

The case where p does not divide n is easily settled. Here $\gcd(p, n) = 1$ and so

$$\phi(pn) = \phi(p)\phi(n) = (p-1)\phi(n).$$

When p does divide n we cannot use the multiplicative property immediately. Let $n = p^k m$, where $\gcd(m, p) = 1$ and $k \geq 1$. Then

$$\phi(pn) = \phi\left(p^{k+1}m\right) = \phi\left(p^{k+1}\right)\phi(m) = p^{k+1}\left(1 - \frac{1}{p}\right)\phi(m)$$

whilst

$$p\phi(n) = p\phi\left(p^k m\right) = p\phi\left(p^k\right)\phi(m) = p \times p^k\left(1 - \frac{1}{p}\right)\phi(m),$$

and the two are seen to be equal.

This result shows that $\phi(n)$ divides $\phi(pn)$ for *any* prime p, regardless of whether or not p is a divisor of n. Now if d is a divisor of n we may write $n = p_1 p_2 \ldots p_r d$, where the p_i are (not necessarily distinct) primes which may or may not divide d. Repeated application of the result gives

$$\phi(d) | \phi(p_1 d), \ \ \phi(p_1 d) | \phi(p_1 p_2 d), \ldots,$$
$$\phi(p_1 p_2 \ldots p_{r-1} d) | \phi(p_1 p_2 \ldots p_{r-1} p_r d),$$

and as the last of these terms is $\phi(n)$ we see that $\phi(d)$ divides $\phi(n)$. ◆

Example 4.2

Find all integers n for which $\phi(n) = 10$.

We first observe that the formula for $\phi(n)$ can be expressed in a slightly different way which removes the reciprocals of the primes.

Alternative formula for $\phi(n)$

If $n = p_1^{k_1} p_2^{k_2} \ldots p_r^{k_r}$, then
$$\phi(n) = p_1^{k_1-1} p_2^{k_2-1} \ldots p_r^{k_r-1}(p_1 - 1)(p_2 - 1)\ldots(p_r - 1).$$

In this formulation some (possibly all) of the exponents $k_i - 1$ may be 0.

The main observation to make from this formula is that if p divides n then $p - 1$ divides $\phi(n)$. For the case $\phi(n) = 10$, the primes p for which $p - 1$ divides 10 are 2, 3 and 11 and so n must be of the form $n = 2^a 3^b 11^c$ for some non-negative integers a, b and c.

Suppose first that $c = 0$. Then $n = 2^a 3^b$ and $\phi(n) = \phi(2^a)\phi(3^b)$. Now $\phi(2^a) = 2^{a-1}$ (for $a \geq 1$) and $\phi(3^b) = 3^{b-1}(3 - 1)$ (for $b \geq 1$) and as neither of these is divisible by 5 there is no solution. So for $\phi(n) = 10$ we must have $c \geq 1$. In this case we can write $n = 11^c m$, where $\gcd(11, m) = 1$. Then

$$\phi(n) = \phi(11^c)\phi(m) = 11^{c-1} \times 10 \times \phi(m) = 10.$$

This only happens when $c = 1$ and $\phi(m) = 1$. Now $\phi(m) = 1$ for $m = 1$ and $m = 2$ alone (see Problem 4.3 below) and so we have just the two integers $n = 11$ and $n = 22$ which satisfy $\phi(n) = 10$. ◆

Problem 4.3 _____

Prove that $\phi(n)$ is even for every integer $n > 2$.

Problem 4.4 _____

Use the formulae obtained for $\sigma(n)$, $\tau(n)$ and $\phi(n)$ to show that if n has prime decomposition $n = p_1^{k_1} p_2^{k_2} \ldots p_r^{k_r}$ then

(a) $\sigma(n)\phi(n) = n^2 \displaystyle\prod_{1 \leq i \leq r} \left(1 - \frac{1}{p_i^{k_i+1}} \right)$;

(b) $\tau(n)\phi(n) \geq n$. *Hint*: Note that $1 - \dfrac{1}{p} \geq \frac{1}{2}$ for any prime p.

Diversion

If $m = 568$ and $n = 638$, then:

$$\sigma(m) = \sigma(n) \quad (= 1080);$$
$$\tau(m) = \tau(n) \quad (= 8);$$
$$\phi(m) = \phi(n) \quad (= 280).$$

In Theorem 1.2 we discovered that whenever f is a multiplicative function then so too is

$$F(n) = \sum_{d|n} f(d).$$

Which function F will result when we take $f = \phi$? To obtain some feeling for F let us work out a sample of its values.

$$F(8) = \phi(1) + \phi(2) + \phi(4) + \phi(8) = 1 + 1 + 2 + 4 = 8$$
$$F(9) = \phi(1) + \phi(3) + \phi(9) = 1 + 2 + 6 = 9$$
$$F(10) = \phi(1) + \phi(2) + \phi(5) + \phi(10) = 1 + 1 + 4 + 4 = 10$$

There are no prizes for spotting the pattern!

Theorem 4.3

$$\sum_{d|n} \phi(d) = n$$

If we try applying any of the various formulae concerning the ϕ-function we shall struggle to make headway towards proving this result. But there is an elegant, snappy proof, first thought of by Gauss. It hinges on the idea of partitioning the integers from 1 to n inclusive into sets in such a way that two integers belong to the same set if, and only if, they have the same greatest common divisor with n. For example, for $n = 12$ we have six classes

$$S_1 = \{1, 5, 7, 11\}, \quad S_2 = \{2, 10\}, \quad S_3 = \{3, 9\},$$
$$S_4 = \{4, 8\}, \qquad S_6 = \{6\}, \qquad S_{12} = \{12\}.$$

The subscript in the label for each class is the greatest common divisor associated with that class; for example, the two members of S_4 are the positive integers not exceeding 12 which have greatest common divisor with 12 equal to 4. The trick now is to form a new set $T_{12/d}$ from S_d by dividing each member of the set by the appropriate greatest common divisor. This gives

$$T_{12} = \{1, 5, 7, 11\}, \quad T_6 = \{1, 5\}, \quad T_4 = \{1, 3\},$$
$$T_3 = \{1, 2\}, \qquad T_2 = \{1\}, \qquad T_1 = \{1\}.$$

For example, S_4 becomes $T_{12/4}$ on dividing each element by 4. Notice that each T_i consists of a reduced set of (least positive) residues for modulus i and so the number of elements in T_i is $\phi(i)$. Adding up over the six classes, the sum of the six values of $\phi(i)$, one for each divisor i of 12, is equal to the total number of integers in the classes, namely 12.

The proof which follows sets this idea out for a general positive integer n.

Proof of Theorem 4.3

Define $S_d = \{m : 1 \leq m \leq n \text{ and } \gcd(m, n) = d\}$. Now $\gcd(m, n) = d$ if, and only if, $\dfrac{m}{d}$ and $\dfrac{n}{d}$ are integers with $\gcd\left(\dfrac{m}{d}, \dfrac{n}{d}\right) = 1$. Therefore the number of integers in S_d is equal to the number of positive integers not exceeding $\dfrac{n}{d}$ which are relatively prime to $\dfrac{n}{d}$, that is, $\phi\left(\dfrac{n}{d}\right)$. As each of the integers $1, 2, \ldots, n$ lies in exactly one class S_d, and there is one class corresponding to each divisor d of n,

$$n = \sum_{d|n} \phi\left(\frac{n}{d}\right) = \sum_{d|n} \phi(d).$$

■ See Problem 1.3.

5 PRIMITIVE ROOTS

5.1 Primitive roots generalized

In Problem 3.4 we saw that the integer 7 had the maximum possible order, $\phi(22)$, modulo 22. When dealing with prime modulus we called an integer with this maximal property a *primitive root*. That definition can be generalized to a composite modulus.

In Section 2.2 of *Unit 4*, we investigated primes p which have 10 as a primitive root.

Definition 5.1 Primitive root

If a has order $\phi(n)$ modulo n then a is said to be a *primitive root of n*.

Remember that the condition $\gcd(a, n) = 1$ has to hold for the order of a to be defined.

Suppose that a is a primitive root of n and consider the set of integers $\{a, a^2, \ldots, a^{\phi(n)}\}$. Each of the integers in this set is relatively prime to n because $\gcd(a, n) = 1$. Also, no two members of this set are congruent modulo n because if $a^r \equiv a^s \pmod{n}$, where $1 \leq s \leq r \leq \phi(n)$, then $a^{r-s} \equiv 1 \pmod{n}$ and Theorem 3.2 informs us that the order of a modulo n, namely $\phi(n)$, divides $r - s$; so $r = s$. As the set contains $\phi(n)$ incongruent integers, this set is a reduced set of residues modulo n. What this is saying is that taking the first $\phi(n)$ powers of any primitive root 'generates' a reduced set of residues.

Theorem 5.1

If a is a primitive root of n then

$$\{a, a^2, a^3, \ldots, a^{\phi(n)}\}$$

is a reduced set of residues modulo n.

For example, in Problem 3.3 we discovered that 3 has order $\phi(14) = 6$ modulo 14, and so is a primitive root of 14. The first six powers of 3 give the members of a reduced set of residues of 14 as follows:

$$3^1 \equiv 3 \pmod{14}; \quad 3^4 \equiv 11 \pmod{14};$$
$$3^2 \equiv 9 \pmod{14}; \quad 3^5 \equiv 5 \pmod{14};$$
$$3^3 \equiv 13 \pmod{14}; \quad 3^6 \equiv 1 \pmod{14}.$$

In Problem 3.3 we also discovered the order of each of these six numbers modulo 14, as summarized in the following table.

Table 5.1 Orders of 3^k modulo 14

k	1	2	3	4	5	6
3^k	3	9	13	11	5	1
Order of 3^k	6	3	2	3	6	1

As you would expect from Theorem 3.2, the order of 3^k always divides $\phi(14) = 6$, but notice that this order is 6 when k is prime to 6, it is 3 when $k = 2$ or 4, it is 2 when $k = 3$ and is 1 when $k = 6$. This suggests the following result.

Theorem 5.2

If a has order c modulo n then, for any $k \geq 1$, a^k has order

$$\frac{c}{\gcd(c, k)} \quad \text{modulo } n.$$

For example, knowing that 3 has order 6 modulo 14, this theorem tells us that:

$$3^2 \text{ has order } \frac{6}{\gcd(6, 2)} = 3; \quad 3^4 \text{ has order } \frac{6}{\gcd(6, 4)} = 3;$$

$$3^3 \text{ has order } \frac{6}{\gcd(6, 3)} = 2; \quad 3^5 \text{ has order } \frac{6}{\gcd(6, 5)} = 6.$$

All of these agree with the final row of Table 5.1.

Proof of Theorem 5.2

Suppose that a has order c modulo n, and consider a^k. Writing $d = \gcd(c, k)$, let $c = dc'$ and $k = dk'$, where $\gcd(c', k') = 1$. Our task is to show that the order of a^k is c'. Let the order of a^k be r. Then

$$(a^k)^{c'} = (a^{dk'})^{c/d} = a^{k'c} = (a^c)^{k'} \equiv (1)^{k'} \equiv 1 \pmod{n},$$

and so, by Theorem 3.2, c' is a multiple of r.

On the other hand

$$a^{kr} = (a^k)^r \equiv 1 \pmod{n},$$

and so kr is a multiple of c. That is, (substituting values for k and c), $dk'r$ is a multiple of dc', which shows that c' divides $k'r$. But then, Euclid's Lemma, which we can apply since $\gcd(c', k') = 1$, tells us that c' divides r.

Putting the two parts together, as c' is both a multiple and a divisor of r, we have that $c' = r$ as required. ∎

As an immediate consequence of this result we can say how many primitive roots n will have and how to find them, provided it has at least one primitive root.

Corollary to Theorem 5.2

If a is a primitive root of n then a^k is also a primitive root of n if, and only if, k is relatively prime to $\phi(n)$.

Furthermore, if n has a primitive root, then it has exactly $\phi(\phi(n))$ primitive roots.

Of course, we count only incongruent primitive roots.

Proof of Corollary

Suppose that a is a primitive root of n. Then $\{a, a^2, \ldots, a^{\phi(n)}\}$ is a reduced set of residues for n and so all primitive roots are contained in this set. Now the order of a^k is $\dfrac{\phi(n)}{\gcd(\phi(n), k)}$ and so a^k will be a primitive root precisely when $\gcd(\phi(n), k) = 1$, that is, when k is relatively prime to $\phi(n)$.

Remember that the primitive roots of n are the integers with order $\phi(n)$.

So the primitive roots are the integers a^k, where k belongs to the set $\{1, 2, \ldots, \phi(n)\}$ and is relatively prime to $\phi(n)$. By definition of ϕ, there are $\phi(\phi(n))$ such exponents k, and hence $\phi(\phi(n))$ primitive roots of n. ∎

Let us put the theorem and its corollary to work.

Example 5.1

Given that 2 is a primitive root of 13 find all the primitive roots of 13.

As 13 allegedly has one primitive root there will be in all $\phi(\phi(13)) = \phi(12) = 4$ primitive roots of 13. The $\phi(12)$ positive integers not exceeding 12 and relatively prime to it are 1, 5, 7 and 11, and, according to the corollary, these are the exponents of 2 which give the primitive roots:

By theorem 5.2, 2^k has order 12 when $\gcd(k, 12) = 1$.

$$2^1 = 2;\ 2^5 = 32 \equiv 6;\ 2^7 \equiv 2^5 \times 4 \equiv 11;\ 2^{11} = 2^5 \times 2^5 \times 2 \equiv 7 \pmod{13},$$

and so the four primitive roots of 13 are 2, 6, 7 and 11. ♦

The following table shows the primitive roots for all moduli up to 10.

Table 5.2   Primitive roots for small integers

n	$\phi(n)$	$\phi(\phi(n))$	Primitive roots
2	1	1	1
3	2	1	2
4	2	1	3
5	4	2	2, 3
6	2	1	5
7	6	2	3, 5
8	4	2	None!
9	6	2	2, 5
10	4	2	3, 7

Note that 8 does not have any primitive roots as $\phi(8) = 4$ and

$$1^2 \equiv 3^2 \equiv 5^2 \equiv 7^2 \equiv 1 \pmod{8}$$

shows that all the integers relatively prime to 8 have order 1 or 2.

Problem 5.1 ⎯⎯⎯⎯⎯⎯⎯⎯⎯⎯⎯⎯⎯⎯⎯⎯⎯⎯⎯⎯⎯

Find all primitive roots, if any, of (a) 11 and (b) 15.

5.2 Moduli admitting primitive roots

Having seen that some moduli have primitive roots and some do not, curiosity leads us to ask the question of just which ones do. To round off this unit we shall give the answer, and show where it comes from, but we shall omit some of the detail. The key result is the following.

> **Theorem 5.3**
>
> Any integer which can be expressed as the product of two divisors which are relatively prime and each exceeding 2 does not have a primitive root.

Proof of Theorem 5.3

Let the integer in question be mn, where $m > 2$, $n > 2$ and $\gcd(m, n) = 1$. Choosing any integer a which is relatively prime to mn, our goal is to show that $a^k \equiv 1 \pmod{mn}$ for some positive integer k less than $\phi(mn)$. In fact $k = \frac{1}{2}\phi(mn)$ will suffice, as we shall show.

We first observe that, since $\phi(m)$ and $\phi(n)$ are both even, (see Problem 4.3), it follows that $\frac{1}{2}\phi(m)$ and $\frac{1}{2}\phi(n)$ are integers.

As $\gcd(a, mn) = 1$ we have that $\gcd(a, m) = 1$, from which Euler's Theorem gives $a^{\phi(m)} \equiv 1 \pmod{m}$. Therefore, with $k = \frac{1}{2}\phi(mn)$,

$$a^k = a^{\phi(m)\phi(n)/2} = (a^{\phi(m)})^{\phi(n)/2} \equiv (1)^{\phi(n)/2} \equiv 1 \pmod{m}.$$

The same argument with the roles of m and n interchanged shows $a^k \equiv 1 \pmod{n}$, and putting the two parts together $a^k \equiv 1 \pmod{mn}$, as required. ∎

Theorem 5.3 may not seem to have made inroads into our immediate problem until you stop and think about which integers cannot be expressed as the product of two relatively prime divisors each exceeding 2. Any number which is divisible by at least two odd primes can be expressed in this way. What does that leave?

> **Corollary to Theorem 5.3**
>
> If an integer n has a primitive root then it must have one of the following forms.
>
> (a) $n = p^k$, p prime, $k \geq 1$
>
> (b) $n = 2p^k$, p an odd prime, $k \geq 1$

Classes (a) and (b) in the Corollary give all the remaining candidates for moduli n which admit a primitive root. But we observe that not all the members of class (a) have the desired property; $8 = 2^3$ has been seen to have no primitive root. In fact all the powers of 2 from 2^3 onwards do not have primitive roots.

> **Theorem 5.4**
>
> For any $k \geq 3$, 2^k does not have a primitive root.

Proof of Theorem 5.4

We know that $\phi(2^k) = 2^{k-1}$, and the 2^{k-1} least positive residues which are relatively prime to 2^k are the odd integers up to $2^k - 1$. So our task is to show that for any odd integer a there exists a positive integer $r < 2^{k-1}$ such that $a^r \equiv 1 \pmod{2^k}$. It turns out that $r = 2^{k-2}$ has precisely this property for every choice of a. To justify this let $P(k)$ be the proposition

$$a^{2^{k-2}} \equiv 1 \pmod{2^k}.$$

$a^{2^{k-2}}$ is $a^{(2^{k-2})}$, not $(a^2)^{k-2}$.

We use Mathematical Induction to prove that $P(k)$ is true for all $k \geq 3$. The case $P(3)$ is a familiar one: $a^2 \equiv 1 \pmod 8$ for all odd integers a. That gives us the basis for induction and so we turn to the induction step. Assume that $P(h)$ is true for some $h \geq 3$. That is, writing the congruence in a more suitable way,

$$a^{2^{h-2}} = 1 + m2^h, \text{ for some integer } m.$$

Squaring both sides

$$\left(a^{2^{h-2}}\right)^2 = \left(1 + m2^h\right)^2$$
$$= 1 + 2\left(m2^h\right) + \left(m2^h\right)^2$$
$$= 1 + 2^{h+1}\left(m + m^2 2^{h-1}\right)$$
$$\equiv 1 \pmod{2^{h+1}}.$$

Hence $a^{2^{h-1}} \equiv 1 \pmod{2^{h+1}}$, confirming the truth of $P(h+1)$ and completing the induction proof. ∎

$\left(a^{2^{h-2}}\right)^2 = a^{2^{h-1}}$

It turns out that the remaining integers listed in the corollary do all have primitive roots and so we have the following.

Theorem 5.5 Integers with primitive roots

The integer n has a primitive root if, and only if, $n = 2$, 4, p^k, $2p^k$, for p an odd prime and $k \geq 1$.

This result, whose proof we shall not complete here, turns out to be of great importance in algebra, in particular in the study of *fields*.

The first proof of this theorem was given by Gauss in 1801. A key stage of the proof is to show that every odd prime has a primitive root. This part was managed in 1773 by Euler, who proved that every prime p has $\phi(p-1)$ primitive roots.

ADDITIONAL EXERCISES

Section 1

1 An integer n is said to be *square-free* if it is not divisible by the square of any prime. Prove that if $n > 1$ is square-free then $\tau(n) = 2^r$ for some $r \geq 1$. Is the converse true, namely that if $\tau(n) = 2^r$ then n must be square-free?

2 Find all integers n for which $\sigma(n) = 24$.

3 Show that $\tau(n)$ is odd if, and only if, n is a square. For which integers n is $\sigma(n)$ odd?

4 Show that $\prod_{d|n} d = n^{\tau(n)/2}$.   *Hint:* If d divides n then $\dfrac{n}{d}$ divides n and $d \times \dfrac{n}{d} = n$.

5 Suppose that the positive integer n has the property that $n + \sigma(n)$ is divisible by 3. Prove the following.

(a) If n is prime then $n \equiv 1 \pmod 6$.

(b) The integer n cannot be the square of a prime.

(c) If $n = pq$, where p and q are distinct odd primes with $p < q$, then $p = 3$ and $q \equiv 5 \pmod 6$.

6 Prove that if f and g are multiplicative functions then the function defined for all positive integers by $F(n) = f(n)g(n)$ is also multiplicative.

Section 2

1 Prove that n is perfect if, and only if,

$$\sum_{d|n} \frac{1}{d} = 2.$$

Deduce that if n is perfect and d is a divisor of n other than n itself, then d is not perfect.

2 Let $n = 2^{k-1}(2^k - 1)$, where the integer $k \geq 3$ is odd. Show that:

(a) $n \equiv 4 \pmod 6$;

(b) $n \equiv 1 \pmod 9$;

(c) n is a triangular number.

These properties hold for all perfect numbers where, in addition, k and $2^k - 1$ are odd primes.

3 An integer $n \geq 1$ is said to be *abundant* if $\sigma(n) > 2n$ and *deficient* if $\sigma(n) < 2n$.

(a) Classify each of the numbers from 24 to 30 inclusive as abundant, deficient or perfect.

(b) Show that if p is prime and $r \geq 1$, then p^r is deficient.

(c) Show that, with one exception, any product of two distinct primes is deficient.

(d) Find an odd abundant number. This is not easy!   *Hint*: You will have to involve three different odd primes.

(e) Show that if m is abundant and $\gcd(m, n) = 1$, then mn is abundant.

(f) Show that for integers $k > 1$, $n = 2^{k-1}(2^k - 1)$ is abundant whenever $2^k - 1$ is not prime.

4 A pair of positive integers m and n which satisfy

$$\sigma(m) = \sigma(n) = m + n$$

is said to be an *amicable pair*.

(a) Verify that the following pairs are amicable:

220 and 284   (Pythagoras, 500 BC);

$17\,296 = 2^4 \times 23 \times 47$ and $18\,416 = 2^4 \times 1151$ (Fermat, 1636);

$2^7 \times 191 \times 383$ and $2^7 \times 73\,727$ (Descartes, 1638).

Note that 1151, 191, 383 and 73 727 are primes.

(b) Show that a prime p cannot be one of an amicable pair.

Section 3

1 (a) Write down a reduced set of residues modulo 20.

(b) Show that $\{5, 5^2, 5^3, 5^4, 5^5, 5^6\}$ is a reduced set of residues modulo 14.

2 Show that $\gcd(a, n) = \gcd(n - a, n)$. Use this to prove that the sum of the integers in the reduced set of least positive residues of n is $\dfrac{n\phi(n)}{2}$.

3 Determine the final two digits of 27^{82} and of 7^{38}.

4 Show that, if $\gcd(a, n) = 1$, then the linear congruence $ax \equiv b \pmod{n}$ has solution $x \equiv ba^{\phi(n)-1} \pmod{n}$. Hence solve the following linear congruences.

(a) $5x \equiv 11 \pmod{12}$

(b) $3x \equiv 7 \pmod{16}$

5 If m and n are relatively prime integers prove that
$$m^{\phi(n)} + n^{\phi(m)} \equiv 1 \pmod{mn}.$$

6 Prove that, for any given positive integer n, there is some positive multiple of n each of whose digits is either 0 or 1.

Section 4

1 Find all solutions of $\phi(n) = 18$.

2 Show that there is no positive integer n for which $\phi(n) = 14$. Confirm that $d = 7$ is the smallest positive integer for which $\phi(n) = 2d$ has no solutions.

3 (a) Show that $\phi(n) = \phi(n + 2)$ is satisfied by $n = 2(2p - 1)$ whenever p and $2p - 1$ are odd primes.

(b) Show that $\phi(n) + 2 = \phi(n + 2)$ is satisfied by $n = 4p$ whenever p and $2p + 1$ are odd primes.

4 Show that $\phi(n^2) = n\phi(n)$, for all integers $n \geq 1$. Hence show that if p is a prime and $k \geq 2$, then
$$\phi(\phi(p^k)) = p^{k-2}\phi((p - 1)^2).$$

Section 5

1 (a) Which integers have order 6

(i) modulo 17,   (ii) modulo 18.

(b) Given that 2 is a primitive root of 19, use Theorem 5.2 to determine which integers have order 6 modulo 19.

2 Let p be an odd prime and suppose that the integer a has order 3 modulo p. Show that
$$1 + a + a^2 \equiv 0 \pmod{p}.$$

Hence prove that $1 + a$ has order 6 modulo p.

3   Use Theorem 5.2 to show that if r is a primitive root of an odd prime p and k is even, then r^k is not a primitive root of p. Hence show that if r and s are primitive roots of some odd prime p then rs is not a primitive root of p.

4   Suppose that the positive integer n has a primitive root. Prove that the product of the numbers which form a reduced set of residues of n is congruent modulo n to -1. Is the result necessarily true if n does not have a primitive root?

Challenge Problems

1   Let n be an odd positive integer. Prove that there are $\tau(n)$ ways of writing n as a sum of consecutive positive integers. For example the $\tau(15) = 4$ ways of writing 15 are

$$1 + 2 + 3 + 4 + 5; \quad 4 + 5 + 6; \quad 7 + 8; \quad 15.$$

How many ways are there of writing an even positive integer as a sum of consecutive positive integers?

2   Prove that n is prime if, and only if, $\phi(n) + \sigma(n) = n\tau(n)$.

3   Prove the following formulae.

(a) $\displaystyle\sum_{d|n} \sigma(d) = \sum_{d|n} \frac{n}{d}\tau(d)$

(b) $\displaystyle\sum_{d|n} \frac{n}{d}\sigma(d) = \sum_{d|n} d\tau(d)$

(c) $d\phi(m)\phi(n) = \phi(mn)\phi(d)$, where m and n are integers with $\gcd(m, n) = d$.

4   Show that an odd perfect number must have at least four prime divisors. *Hint:* Suppose that $n = p^a q^b r^c$ is perfect. Show that

$$\sigma(n) < \frac{npqr}{(p-1)(q-1)(r-1)}$$

and deduce that for n to be perfect

$$\left(1 - \frac{1}{p}\right)\left(1 - \frac{1}{q}\right)\left(1 - \frac{1}{r}\right) < \frac{1}{2}.$$

Show that this is impossible.

5   Let p be an odd prime. Show that:

(a) if a is a primitive root of p then either a or $a + p$ is a primitive root of p^2;

(b) if a is a primitive root of p^2 then a is a primitive root of p^k for any $k \geq 1$;

(c) if a is a primitive root of p^k then, if a is odd it is a primitive root of $2p^k$, and if a is even then $a + p^k$ is is a primitive root of $2p^k$.

SOLUTIONS TO THE PROBLEMS

Solution 1.1

n	1	2	3	4	5	6	7	8	9	10
$\tau(n)$	1	2	2	3	2	4	2	4	3	4
$\sigma(n)$	1	3	4	7	6	12	8	15	13	18

Solution 1.2

The divisors of 8 are 1, 2, 4, and 8 itself. Therefore

$$\sum_{d|8} \sigma(d) = \sigma(1) + \sigma(2) + \sigma(4) + \sigma(8)$$
$$= 1 + 3 + 7 + 15 = 26,$$

and

$$\prod_{d|8} \tau(d) = \tau(1) \times \tau(2) \times \tau(4) \times \tau(8)$$
$$= 1 \times 2 \times 3 \times 4 = 24.$$

Solution 1.3

If d is a divisor of n then $d \times \dfrac{n}{d} = n$ and so $\dfrac{n}{d}$ is also a divisor of n.

Therefore if $d_1, d_2, \ldots, d_r$ are the distinct divisors of n then $\dfrac{n}{d_1}, \dfrac{n}{d_2}, \ldots, \dfrac{n}{d_r}$ is the same collection of divisors, but in a different order. It follows that

$$\sum_{d|n} f\left(\frac{n}{d}\right) = f\left(\frac{n}{d_1}\right) + f\left(\frac{n}{d_2}\right) + \cdots + f\left(\frac{n}{d_r}\right)$$
$$= f(d_1) + f(d_2) + \cdots + f(d_r)$$
$$= \sum_{d|n} f(d).$$

Applying this formula in the case where f is the reciprocal function we have

$$\sum_{d|n} \frac{1}{d} = \sum_{d|n} \frac{1}{\left(\frac{n}{d}\right)} = \frac{1}{n}\left(\sum_{d|n} d\right) = \frac{\sigma(n)}{n}.$$

Solution 1.4

(a) $360 = 2^3 \times 3^2 \times 5$ and so:

$$\tau(360) = \tau(2^3) \times \tau(3^2) \times \tau(5) = 4 \times 3 \times 2 = 24;$$
$$\sigma(360) = \sigma(2^3) \times \sigma(3^2) \times \sigma(5)$$
$$= (1 + 2 + 4 + 8)(1 + 3 + 9)(1 + 5) = 15 \times 13 \times 6 = 1170.$$

(b) $$\tau(2p^3) = \tau(2)\tau(p^3) = 2 \times 4 = 8;$$
$$\sigma(3p) = \sigma(3)\sigma(p) = 4(p + 1);$$
$$\sigma(6p^2) = \sigma(2)\sigma(3)\sigma(p^2) = 3 \times 4 \times (p^2 + p + 1) = 12(p^2 + p + 1).$$

Solution 1.5

From the multiplicative property $f(mn) = f(m)f(n)$, the case $n = 1$ gives $f(m) = f(m)f(1)$. For this to hold we either have $f(1) = 1$ or $f(m) = 0$ for all positive integers m. In the latter case f is the zero function which is trivially multiplicative. With the exception of the zero function, all multiplicative functions f have $f(1) = 1$.

Solution 1.6

Let m and n be any two integers. If either of these is even, then

$$f(mn) = 0 = f(m)f(n).$$

On the other hand, if m and n are both odd, then

$$f(mn) = mn = f(m)f(n).$$

Since these cover all cases, it follows that f is a multiplicative function.

$$F(n) = \sum_{d|n} f(d) = \text{the sum of the } odd \text{ divisors of } n.$$

The condition that $\gcd(m, n) = 1$ is not needed.

Each even divisor of n contributes 0 to the sum.

Solution 1.7

To apply Theorem 1.2 we require a multiplicative function f such that

$$\sigma_r(n) = \sum_{d|n} d^r = \sum_{d|n} f(d).$$

It follows that f must be the function defined by $f(n) = n^r$, and for this function,

$$f(mn) = (mn)^r = m^r n^r = f(m)f(n).$$

So f, and with it σ_r, is a multiplicative function.

Solution 2.1

Suppose that $2^k - 1 = p$, a prime. Then p is odd and

$$\sigma(p) = p + 1 = 2^k.$$

As $\gcd(2^{k-1}, p) = 1$, the multiplicative property of σ gives

$$\sigma(n) = \sigma(2^{k-1}p) = \sigma(2^{k-1})\sigma(p) = (2^k - 1)2^k$$
$$= 2(2^{k-1}(2^k - 1)) = 2n.$$

$\sigma(2^r) = 2^{r+1} - 1$

Thus n is perfect.

Solution 2.2

Since q is prime and $\gcd(2, q) = 1$, FLT gives $2^{q-1} \equiv 1 \pmod{q}$, that is,

$$2^{2p} - 1 \equiv 0 \pmod{q}.$$

Factorizing the left-hand side of this congruence gives

$$(2^p - 1)(2^p + 1) \equiv 0 \pmod{q}.$$

As q is prime, Euclid's Lemma gives

$$q|(2^p - 1) \text{ or } q|(2^p + 1),$$

which is the claimed result as $M_p = 2^p - 1$.

Solution 2.3

The square root of M_{17} is approximately $2^{17/2}$ which is close to 362. So if M_{17} is composite it must have a prime divisor of the form $34k + 1$ which is less than 362. Of the ten numbers of this form which are less than 362, namely

$$35, \ 69, \ \underline{103}, \ \underline{137}, \ 171, \ 205, \ \underline{239}, \ 273, \ \underline{307}, \ 341,$$

just the four numbers underlined are prime. Either M_{17} is prime or it is divisible by one of these four primes.

$2^{17/2} = 2^8 \sqrt{2} = 256 \times \sqrt{2} \simeq 362.04$

Solution 3.1

n	Reduced set of residues	$\phi(n)$
1	$\{1\}$	1
2	$\{1\}$	1
3	$\{1,2\}$	2
4	$\{1,3\}$	2
5	$\{1,2,3,4\}$	4
6	$\{1,5\}$	2
7	$\{1,2,3,4,5,6\}$	6
8	$\{1,3,5,7\}$	4
9	$\{1,2,4,5,7,8\}$	6
10	$\{1,3,7,9\}$	4

Solution 3.2

Of the integers in the set

$$\{\underline{1}, 2, \underline{3}, 4, \underline{5}, 6, 7, 8, \underline{9}, 10, \underline{11}, 12, \underline{13}\}$$

only the six underlined integers are relatively prime to 14, so $\phi(14) = 6$.

$$3^6 = (3^3)^2 = (27)^2 \equiv (-1)^2 \equiv 1 \ (\text{mod } 14)$$
$$5^6 = (5^2)^3 = 25^3 \equiv (-3)^3 \equiv -27 \equiv 1 \ (\text{mod } 14)$$

Solution 3.3

The least positive residues in a reduced set modulo 14 are 1, 3, 5, 9, 11 and 13.

$1^1 \equiv 1 \ (\text{mod } 14)$, so 1 has order 1.
$3^2 \equiv 9$, $3^3 \equiv 13$, $3^4 \equiv 11$, $3^5 \equiv 5$, $3^6 \equiv 1 \ (\text{mod } 14)$, so 3 has order 6.
$5^2 \equiv 11$, $5^3 \equiv 13$, $5^4 \equiv 9$, $5^5 \equiv 3$, $5^6 \equiv 1 \ (\text{mod } 14)$, so 5 has order 6.
$9^2 \equiv 11$, $9^3 \equiv 1 \ (\text{mod } 14)$, so 9 has order 3.
$11^2 \equiv 9$, $11^3 \equiv 1 \ (\text{mod } 14)$, so 11 has order 3
$13^2 \equiv (-1)^2 \equiv 1 \ (\text{mod } 14)$, so 13 has order 2.

Solution 3.4

(a) The prime decomposition of 21 is 3×7, so $\phi(21) = 12$, as can be seen by counting how many integers in $\{1, 2, \ldots, 19, 20\}$ are not divisible by either 3 or 7. So the possible orders of 4 modulo 21 are 1, 2, 3, 4, 6 and 12.

$$4^1 \equiv 4; \quad 4^2 \equiv 16; \quad 4^3 = 64 \equiv 1 \ (\text{mod } 21),$$

and so 4 has order 3.

(b) $\phi(22) = 10$ and so the possible orders of 7 modulo 22 are 1, 2, 5 and 10.

$$7^1 \equiv 7; \quad 7^2 = 49 \equiv 5;$$
$$7^5 = 7^2 \times 7^2 \times 7 \equiv 5 \times 5 \times 7 \equiv 3 \times 7 \equiv 21 \ (\text{mod } 22).$$

As 7 does not have one of the smaller orders it must have order 10. Of course, $7^{10} \equiv 1 \ (\text{mod } 22)$ by Euler's Theorem.

Solution 3.5

As $100 = 2^2 5^2$, $\phi(100)$ is equal to the number of integers in the set $\{1, 2, 3, \ldots, 99, 100\}$ which are not multiples of either 2 or 5, that is, those numbers which end in 1, 3, 7 or 9. There are 40 such numbers and so $\phi(100) = 40$.

The final two digits of a number are determined by its least positive residue modulo 100.

(a) By Euler's Theorem, $17^{40} \equiv 1 \pmod{100}$, and so

$$17^{83} \equiv (17^{40})^2 17^3 \equiv 17^3 \equiv 289 \times 17 \equiv (-11) \times 17 \equiv -187 \equiv 13 \pmod{100},$$

and so the final two digits of 17^{83} are 13.

(b) Since $19^{40} \equiv 1 \pmod{100}$, 19^{39} is a solution of $19x \equiv 1 \pmod{100}$.

We can simplify this congruence by multiplying through by any integer which is relatively prime to 100:

$$19x \equiv 1 \pmod{100} \iff 209x \equiv 11 \pmod{100}$$
$$\iff 9x \equiv 11 \pmod{100}.$$

We can replace the right-hand side by a congruent number, so taking advantage of the result that a number is divisible by 9 when the sum of its digits is divisible by 9:

$$19x \equiv 1 \pmod{100} \iff 9x \equiv 711 \pmod{100}$$
$$\iff x \equiv 79 \pmod{100},$$

and the final two digits of 19^{39} are 79.

Solution 3.6

Let $P(m)$ be the statement that R_{mn} is divisible by R_n.

$P(1)$ is certainly true as it asserts that R_n is divisible by R_n, so we have the basis for induction.

Turning to the induction step, suppose that $P(k)$ is true and consider $R_{(k+1)n}$.

$$R_{(k+1)n} = R_{kn+n} = R_{kn} \times 10^n + R_n.$$

By the induction hypothesis R_n divides R_{kn} and so divides both terms on the right-hand side of this equation. It follows that R_n divides the left-hand side, $R_{(k+1)n}$, as required.

Hence by induction, R_{mn} is divisible by R_n, for all integers $m \geq 1$.

Solution 3.7

(a) Euler's Theorem gives $10^{p-1} \equiv 1 \pmod{p}$. In other words p divides $10^{p-1} - 1$, which amounts to p divides $9R_{p-1}$. However $\gcd(9, p) = 1$, and so p divides R_{p-1}.

Remember that $p > 5$. The result is not true for $p = 2$, $p = 3$ or $p = 5$.

(b) Suppose that n does not divide $p - 1$. Then $p - 1 = nq + r$, where $0 < r < n$.

We are given that p divides R_n (and therefore divides R_{nq} by Problem 3.6). Moreover we know that p divides R_{nq+r} by part (a). If we now look at

$$R_{nq+r} = R_{nq} \times 10^r + R_r$$

we see that p divides two of the terms in this equation and so must divide the third. That is, p divides R_r, contradicting the fact that R_n is the smallest repunit which is a multiple of p. Thus $p - 1$ is a multiple of n, as required.

(c) (i) We know that 7 divides R_6, and by part (b) the only smaller repunits which may be divisible by 7 are $R_2 = 11$ and $R_3 = 111$. As neither of these is divisible by 7, R_6 is the smallest such repunit.

(ii) 13 divides R_{12}, and the candidates for smaller multiples of 13 are R_2, R_3, R_4 and R_6. In fact 13 divides $R_6 = 111\,111$ but not R_2, R_3 or R_4. So the smallest repunit divisible by 13 is R_6.

Solution 3.8

The modulus is the product of relatively prime divisors 3, 5, 16 and 17. For the primes 3, 5 and 17, FLT gives, respectively, $n^3 \equiv n \pmod{3}$, $n^5 \equiv n \pmod{5}$ and $n^{17} \equiv n \pmod{17}$, for all integers n. Therefore

$$n^{33} = (n^3)^{11} \equiv n^{11} \equiv (n^3)^3 n^2 \equiv n^3 n^2 \equiv n^3 \equiv n \pmod{3},$$

$$n^{33} = (n^5)^6 n^3 \equiv n^6 n^3 \equiv n^5 n^4 \equiv n n^4 \equiv n^5 \equiv n \pmod{5},$$

$$n^{33} = n^{17} n^{16} \equiv n n^{16} \equiv n^{17} \equiv n \pmod{17}.$$

For modulus 16 we must use Euler's Theorem. As every odd integer is relatively prime to 16 while every even integer is not, it follows that $\phi(16) = 8$. Therefore, for all odd integers n, $n^8 \equiv 1 \pmod{16}$ and so

$$n^{33} = n(n^8)^4 \equiv n(1)^4 \equiv n \pmod{16}.$$

All four congruences hold for any odd integer n and so, for such n, $n^{33} \equiv n \pmod{15 \times 16 \times 17}$ follows.

Note that FLT must be used in this form because the result is needed for all odd integers, so cases with n congruent to 0 must not be overlooked. Handling the 15 via Euler's Theorem, with $n^{\phi(15)} \equiv 1 \pmod{15}$, will not cater for all cases.

Solution 4.1

$$\phi(24) = \phi(3 \times 8) = \phi(3)\phi(8) = 2 \times 4 = 8$$

$$\phi(70) = \phi(2 \times 5 \times 7) = \phi(2)\phi(5)\phi(7) = 1 \times 4 \times 6 = 24$$

$$\phi(420) = \phi(3 \times 4 \times 5 \times 7) = \phi(3)\phi(4)\phi(5)\phi(7) = 2 \times 2 \times 4 \times 6 = 96$$

Solution 4.2

$$\phi(336) = \phi(2^4)\phi(3)\phi(7)$$

$$= 2^4 \times 3 \times 7 \times \frac{1}{2} \times \frac{2}{3} \times \frac{6}{7} = 96$$

$$\phi(8p^3) = \phi(8)\phi(p^3) = 4p^3 \left(1 - \frac{1}{p}\right) = 4p^2(p-1)$$

Solution 4.3

If n is divisible by an odd prime p then $\phi(n)$ is divisible by $p - 1$ and consequently is even. If n is not divisible by an odd prime then $n = 2^k$ and $\phi(n) = 2^{k-1}$, which is even for $k > 1$. The only numbers not covered are $n = 1$ and $n = 2$; in both these cases $\phi(n) = 1$.

Solution 4.4

(a) From the formulae for σ and ϕ:

$$\sigma(n)\phi(n) = \frac{p_1^{k_1+1} - 1}{p_1 - 1} \ldots \frac{p_r^{k_r+1} - 1}{p_r - 1} \times p_1^{k_1-1} \ldots p_r^{k_r-1}(p_1 - 1) \ldots (p_r - 1)$$

$$= (p_1^{k_1+1} - 1) \ldots (p_r^{k_r+1} - 1)p_1^{k_1-1} \ldots p_r^{k_r-1}$$

$$= p_1^{2k_1} \ldots p_r^{2k_r} \left(1 - \frac{1}{p_1^{k_1+1}}\right) \ldots \left(1 - \frac{1}{p_r^{k_r+1}}\right)$$

$$= n^2 \prod_{1 \le i \le r} \left(1 - \frac{1}{p_i^{k_i+1}}\right).$$

(b) Since $\left(1 - \frac{1}{p}\right) \ge \frac{1}{2}$ for each prime p, we note that

$$\phi(n) = n \prod_{1 \le i \le r} \left(1 - \frac{1}{p_i}\right) \ge \frac{n}{2^r}.$$

Now $\tau(n) = (k_1 + 1)(k_2 + 1) \ldots (k_r + 1) \ge 2^r$, since each $k_i \ge 1$.

From the two inequalities $\phi(n)\tau(n) \ge n$ follows.

Solution 5.1

(a) As $\phi(\phi(11)) = \phi(10) = 4$, there are either none or four primitive roots of 11. In fact 2 is seen to be a primitive root since its powers modulo 11 begin

$$2, \ 2^2 = 4, \ 2^3 = 8, \ 2^4 \equiv 5 \text{ and } 2^5 \equiv 10 \ (\text{mod } 11).$$

Therefore the order of 2 exceeds 5, and as it divides $\phi(11) = 10$, it must actually be 10.

The four primitive roots of 11 are therefore 2^1, 2^3, 2^7 and 2^9, where the exponents $\{1, 3, 7, 9\}$ form a reduced set of residues for $\phi(11) = 10$. Simplifying these values modulo 11 we get

$$2, \ 8, \ 2^7 = 2^4 \times 2^3 \equiv 5 \times 8 \equiv 7 \text{ and } 2^9 = 2^5 \times 2^4 \equiv 10 \times 5 \equiv 6.$$

(b) $\phi(\phi(15)) = \phi(8) = 4$ and so there are either no primitive roots of 15 or four of them. The reduced set of least positive residues modulo 15 is $\{1, 2, 4, 7, 8, 11, 13, 14\}$ and the primitive roots, that is, integers of order 8, are to be found in this set. Now $2^4 \equiv 1 \ (\text{mod } 15)$ and so 2 is not a primitive root as it has order 4. Similarly, $4^2 \equiv 1 \ (\text{mod } 15)$, $14^2 \equiv (-1)^2 \equiv 1 \ (\text{mod } 15)$ and $11^2 \equiv (-4)^2 \equiv 1 \ (\text{mod } 15)$ show that none of 4, 11 or 14 is a primitive root. As 1 always has order 1 we are left with just three candidates for primitive roots, namely 7, 8 and 13. As there has to be either none or four in all, we conclude that 15 has no primitive roots.

SOLUTIONS TO ADDITIONAL EXERCISES

Section 1

1 If n is square-free then each exponent in the prime decomposition of n is less than 2; that is, $n = p_1 p_2 \ldots p_r$ is a product of distinct primes and

$$\tau(n) = 2 \times 2 \times 2 \times \cdots \times 2 = 2^r,$$

where r is the number of primes involved.

The converse is not true. For example, $\tau(2^3) = 4$, and 2^3 is certainly not square-free.

2 Knowing that $\sigma(n) > n$, one approach is to determine the value of $\sigma(n)$ for each n up to and including 23; all the integers n for which $\sigma(n) = 24$ will turn up in this list. But to be a little more sophisticated let us make use of the formula

$$\sigma(p_1^{k_1} p_2^{k_2} \ldots p_r^{k_r}) = (1 + p_1 + p_1^2 + \cdots + p_1^{k_1}) \ldots (1 + p_r + p_r^2 + \cdots + p_r^{k_r}).$$

If n has only one prime divisor then

$$\sigma\left(p_1^{k_1}\right) = \left(1 + p_1 + p_1^2 + \cdots + p_1^{k_1}\right) = 24.$$

This is not solvable for $p_1 = 2$ or 3 and when $p_1 \geq 5$ we have $1 + p_1 + p_1^2 > 30$. That leaves $1 + p_1 = 24$ which has solution $p_1 = 23$, and $n = 23$.

If n has more than one prime divisor we can appeal to its multiplicative nature and from $n = rs$ with $r \geq 2$, $s \geq 2$ and $\gcd(r, s) = 1$, we have $\sigma(n) = \sigma(r)\sigma(s) = 24$. Now the least value that $\sigma(r)$ can take is 3 and the only ways that 24 can occur as the product of divisors not less than 3 is as 3×8 or 4×6.

The values of $\sigma(n)$ which do not exceed 10 are listed in the solutions to Problem 1.1.

$\sigma(r) = 3$ and $\sigma(s) = 8$ imply $r = 2$ and $s = 7$ giving $n = 14$.

$\sigma(r) = 4$ and $\sigma(s) = 6$ imply $r = 3$ and $s = 5$ giving $n = 15$.

Hence $\sigma(n) = 24$ for $n = 14$, 15 and 23.

3   From the formula $\tau(n) = (k_1 + 1)(k_2 + 1)\ldots(k_r + 1)$, where the k_i are the exponents in the prime decomposition of n, we see that $\tau(n)$ is odd if, and only if, each exponent is even. Furthermore, each exponent is even if, and only if, n is a square.

Now $\sigma(n)$ is a product of terms $(1 + p + p^2 + \cdots + p^k)$, one for each prime divisor of n. For $\sigma(n)$ to be odd each of these terms must be odd. When $p = 2$ the term is odd for all values of k. When p is odd $1 + p + p^2 + \cdots + p^k$ is odd only when k is even; so the exponent of each odd prime must be even. We conclude that $\sigma(n)$ is odd when n is any power of 2 (possibly 2^0) times the square of an odd number, that is, $\sigma(n)$ is odd when n is either a square or twice a square.

4   If d is a divisor of n then it has a complementary divisor $d' = \dfrac{n}{d}$ such that $dd' = n$. Now $d \neq d'$ except for the case when n is a square and $d = \sqrt{n}$. So, when n is not a square the $\tau(n)$ divisors partition into $\tau(n)/2$ pairs, each pair having product n, and the product of all the divisors is $n^{\tau(n)/2}$. On the other hand, when n is a square the $\tau(n)$ divisors partition into $(\tau(n) - 1)/2$ pairs, each pair having product n, the remaining divisor being $\sqrt{n}$. This time the product of all the divisors is equal to $n^{(\tau(n)-1)/2} \times n^{1/2}$, which again is $n^{\tau(n)/2}$.

5   (a) If n is the prime p, then $n + \sigma(n) = p + (1 + p) = 2p + 1$. For this to be divisible by 3 we require $2p + 1 \equiv 0 \pmod 3$. This has solution $p \equiv 1 \pmod 3$. Advancing to modulus 6,

$$p \equiv 1 \pmod 3 \implies p \equiv 1, 4 \pmod 6,$$

but numbers congruent to 4 modulo 6 are divisible by 2 and so cannot be prime. We conclude that, in this case, $p \equiv 1 \pmod 6$.

(b) If $n = p^2$, for some prime p, then

$$n + \sigma(n) = p^2 + (p^2 + p + 1) = 2p^2 + p + 1.$$

Now when $p = 3$ this quadratic has value 22 which is not divisible by 3. So $p \neq 3$, in which case $p^2 \equiv 1 \pmod 3$ and

$$2p^2 + p + 1 \equiv p \not\equiv 0 \pmod 3.$$

Hence $p^2 + \sigma(p^2)$ is not divisible by 3.

(c) If $n = pq$ then

$$n + \sigma(n) = pq + (pq + p + q + 1) = 2pq + p + q + 1.$$

Working modulo 3, p can take any of the values 0, 1 or 2 but q can take only values 1 or 2 since $q > 3$ is prime.

The following table gives the possible values modulo 3 of $2pq + p + q + 1$.

	$p \equiv 0$	$p \equiv 1$	$p \equiv 2$
$q \equiv 1$	2	2	2
$q \equiv 2$	0	2	1

The only 0 in the table, corresponding to divisibility by 3, occurs when $p \equiv 0 \pmod 3$ and $q \equiv 2 \pmod 3$, that is, $p = 3$ and $q \equiv 2 \pmod 3$. Now just as in part (a)

$$q \equiv 2 \pmod 3 \implies q \equiv 2 \text{ or } 5 \pmod 6,$$

and as the former value is impossible for an odd prime, we have $q \equiv 5 \pmod 6$.

6   Let m and n be relatively prime integers. Then

$$
\begin{aligned}
F(mn) &= f(mn)g(mn), && \text{by the definition of } F, \\
&= (f(m)f(n))\,(g(m)g(n)), && \text{by the multiplicative property of } f \text{ and } g, \\
&= (f(m)g(m))\,(f(n)g(n)), && \text{by rearranging}, \\
&= F(m)F(n).
\end{aligned}
$$

This confirms that F is multiplicative.

Section 2

1   Suppose that n is perfect. Then $\sum_{d|n} d = 2n$. Using Problem 1.3,

$$2n = \sum_{d|n} d = \sum_{d|n} \frac{n}{d} = n \left(\sum_{d|n} \frac{1}{d} \right),$$

from which $\sum_{d|n} \frac{1}{d} = 2$ follows.

The argument reverses. If $\sum_{d|n} \frac{1}{d} = 2$ then $\sum_{d|n} \frac{n}{d} = 2n$, from which we conclude that n is perfect.

If $d < n$ is a divisor of n then $\sum_{r|d} \frac{1}{r} < \sum_{r|n} \frac{1}{r}$, since each divisor of d is also a divisor of n but at least one divisor of n (namely n itself) is not a divisor of d. If n is perfect the sum on the right of the inequality is equal to 2 and so the sum on the left must be less than 2. Hence d is not perfect.

2   (a) Working modulo 6 the successive powers of 2 alternate 2, 4, 2, 4, ..., with value 2 when the exponent is odd. As k is odd,

$$n = 2^{k-1}(2^k - 1) \equiv 4(2 - 1) \equiv 4 \pmod 6.$$

     (b) Working modulo 9 the successive powers of 2 cycle 2, 4, 8, 7, 5, 1, 2, 4, 8, ... As k is odd 2^{k-1} is congruent modulo 9 to one to 4, 7, or 1 giving, respectively, the following possibilities for n:

$$n = 2^{k-1}(2^k - 1) \equiv 4(8 - 1) \equiv 28 \equiv 1 \pmod 9;$$
$$n = 2^{k-1}(2^k - 1) \equiv 7(5 - 1) \equiv 28 \equiv 1 \pmod 9;$$
$$n = 2^{k-1}(2^k - 1) \equiv 1(2 - 1) \equiv 1 \pmod 9.$$

In all cases n is congruent modulo 9 to 1.

     (c) $n = \frac{1}{2} \times 2^k(2^k - 1) = \frac{1}{2} \times m(m + 1)$, where $m = 2^k - 1$. This is the triangular number T_m.

3   (a) $\sigma(24) = 60 > 48$, so 24 is abundant.
        $\sigma(25) = 31 < 50$, so 25 is deficient.
        $\sigma(26) = 42 < 52$, so 26 is deficient.
        $\sigma(27) = 40 < 54$, so 27 is deficient.
        $\sigma(28) = 56 = 56$, so 28 is perfect.
        $\sigma(29) = 30 < 58$, so 29 is deficient.
        $\sigma(30) = 72 > 60$, so 30 is abundant.

(b) $\sigma(p^r) = 1 + p + p^2 + \cdots + p^{r-1} + p^r < 2p^r$ since

$$1 + p + p^2 + \cdots + p^{r-1} = \frac{p^r - 1}{p - 1} < p^r.$$

The required inequality can also be proved using Mathematical Induction.

So p^r is deficient.

(c) Let p and q be distinct primes with $p < q$. Then

$$\sigma(pq) = \sigma(p)\sigma(q) = (p+1)(q+1) = pq + p + q + 1.$$

This will be less than $2pq$ whenever $p + q + 1 < pq$. Now

$$p + q + 1 \leq 2q \quad \text{(equality holding only when } q = p+1; \text{ that is, when } p = 2, q = 3)$$
$$\leq pq \quad \text{(equality holding only when } p = 2).$$

Hence any number of the form pq is deficient except for the single case $pq = 6$ which is perfect.

(d) The smallest odd abundant number is $n = 3^3 \times 5 \times 7$ which has $\sigma(n) = 40 \times 6 \times 8 = 1920 > 1890 = 2n$.

(e) $\sigma(mn) = \sigma(m)\sigma(n) > 2mn$, since $\sigma(m) > 2m$ and $\sigma(n) \geq n$.

(f) Since $\gcd(2^{k-1}, 2^k - 1) = 1$,

$$\sigma\big(2^{k-1}(2^k - 1)\big) = \sigma(2^{k-1})\,\sigma(2^k - 1) = (2^k - 1)\,\sigma(2^k - 1).$$

Now, for any $m > 1$, $\sigma(m) \geq m + 1$, with equality holding only when m is prime. Hence, if $2^k - 1$ is not prime, then $\sigma(2^k - 1) > 2^k$, and

$$\sigma\big(2^{k-1}(2^k - 1)\big) > 2 \times 2^{k-1}(2^k - 1).$$

This shows that $2^{k-1}(2^k - 1)$ is abundant.

4   (a) $\sigma(220) = \sigma(2^2 \times 5 \times 11) = 7 \times 6 \times 12 = 504 = 220 + 284$;
      $\sigma(284) = \sigma(2^2 \times 71) = 7 \times 72 = 504 = 220 + 284$.

      $\sigma(2^4 \times 23 \times 47) = 31 \times 24 \times 48 = 35\,712 = 17\,296 + 18\,416$;
      $\sigma(2^4 \times 1151) = 31 \times 1152 = 35\,712 = 17\,296 + 18\,416$.

      $\sigma(2^7 \times 191 \times 383) = 255 \times 192 \times 384 = 18\,800\,640$;
      $\sigma(2^7 \times 73\,727) = 255 \times 73\,728 = 18\,800\,640$;
      $2^7 \times 191 \times 383 + 2^7 \times 73\,727 = 9\,363\,584 + 9\,437\,056 = 18\,800\,640$.

So in each case the pair is amicable.

(b) $\sigma(p) = p + 1$, and so p can be amicable only with 1. But $\sigma(1) = 1 \neq p + 1$.

Section 3

1   (a) Those integers between 1 and 20 inclusive which are relatively prime to 20 form the reduced set of residues $\{1, 3, 7, 9, 11, 13, 17, 19\}$.

(b) As $\phi(14) = 6$ and $\gcd(5, 14) = 1$, the set $\{5, 5^2, 5^3, 5^4, 5^5, 5^6\}$ will be a reduced set of residues provided they are distinct modulo 14. This is the case since

$$5^2 \equiv 11, \ 5^3 = 125 \equiv 13, \ 5^4 \equiv 5 \times 13 \equiv 9,$$
$$5^5 \equiv 5 \times 9 \equiv 3, \ 5^6 \equiv 5 \times 3 \equiv 1.$$

2   In establishing the Euclidean Algorithm in Section 5.2 of *Unit 1* we had the result

$$\gcd(a, b) = \gcd(a - qb, b), \quad \text{for any integer } q.$$

In *Unit 1* we used r in place of $a - qb$.

Applied here:

$$\gcd(n - a, n) = \gcd(-a, n) = \gcd(a, n).$$

The result is readily verified for the case $n = 2$ and so we shall assume that $n \geq 3$. Consider the set of integers from 1 to $n - 1$ inclusive which are relatively prime to n. By the above, these break into pairs, a and $n - a$, which have sum n. If $a = n - a$ then $2a = n$, so a is a divisor of n and $\gcd(a, n) = a = \dfrac{n}{2} > 1$, and so a is not relatively prime to n. Hence the $\phi(n)$ integers break into $\phi(n)/2$ pairs of distinct integers with sum n, and so the total sum is $n\phi(n)/2$.

3   The final two digits of an integer are given by its least positive residue modulo 100. Now $\phi(100) = 40$ (see Problem 3.5) and so $27^{40} \equiv 1 \pmod{100}$. Therefore

$$27^{82} = (27^{40})^2 27^2 \equiv 27^2 \equiv 729 \equiv 29 \pmod{100},$$

and so the final two digits are 29.

Since $7^{40} \equiv 1 \pmod{100}$, 7^{38} is a solution of $7^2 x \equiv 1 \pmod{100}$. However,

$$
\begin{aligned}
7^2 x \equiv 1 \pmod{100} &\iff 49x \equiv 1 \pmod{100} \\
&\iff 49x \equiv 301 \pmod{100} \\
&\iff 7x \equiv 43 \pmod{100} \\
&\iff 7x \equiv 343 \pmod{100} \\
&\iff x \equiv 49 \pmod{100},
\end{aligned}
$$

and the required two digits are 49.

4   If $x \equiv ba^{\phi(n)-1} \pmod{n}$ then $ax \equiv ba^{\phi(n)} \pmod{n}$, and since Euler's Theorem gives $a^{\phi(n)} \equiv 1 \pmod{n}$, we have $ax \equiv b \pmod{n}$. As $\gcd(a, n) = 1$ this is the unique solution modulo n, by Theorem 3.2 of *Unit 3*.

Applying this result to the given congruences and noting that $\phi(12) = 4$ and $\phi(16) = 8$:

$$5x \equiv 11 \pmod{12} \text{ has solution } x \equiv 11 \times 5^3 \equiv 55 \times 25 \equiv 55 \equiv 7 \pmod{12};$$
$$3x \equiv 7 \pmod{16} \text{ has solution } x \equiv 7 \times 3^7 \equiv 7 \times 81 \times 27 \equiv 7 \times 1 \times 11 \equiv 13 \pmod{16}.$$

5   If m and n are relatively prime then, by Euler's Theorem, $m^{\phi(n)} \equiv 1 \pmod{n}$ and $n^{\phi(m)} \equiv 1 \pmod{m}$. Therefore

$$m^{\phi(n)} + n^{\phi(m)} \equiv 0 + 1 = 1 \pmod{m}, \text{ and}$$
$$m^{\phi(n)} + n^{\phi(m)} \equiv 1 + 0 = 1 \pmod{n}.$$

As $\gcd(m, n) = 1$ it follows from the corollary to Theorem 1.3 of *Unit 3* that $m^{\phi(n)} + n^{\phi(m)} \equiv 1 \pmod{mn}$.

6   Suppose that $n = 2^r 5^s m$, where either of r and s may be 0 and $\gcd(m, 10) = 1$. We have seen that there is some multiple of m, say km, which is a repunit, that is, all the digits of km are 1's. Hence, if we can find a further multiple which equalizes the exponents of 2 and 5 so that we are multiplying km by some number of 10's, the resulting number will be the repunit km with a string of 0's appended.

Let $t = \max\{r, s\}$. Then $(2^{t-r} 5^{t-s} k) n = 10^t km$ is a suitable multiple.

Section 4

1 If a prime p is a divisor of n then $p - 1$ is necessarily a divisor of $\phi(n)$. The primes p for which $p - 1$ is a divisor of 18 are 2, 3, 7 and 19.

Suppose that 19 divides n. Clearly 19^2 cannot divide n because $\phi(19^2)$ is well in excess of 18. So $n = 19m$, where $\gcd(19, m) = 1$ and

$$\phi(n) = \phi(19)\phi(m) = 18\phi(m) = 18.$$

Now $\phi(m) = 1$ only for $m = 1$ and 2, so we obtain two solutions $n = 19$ and $n = 38$.

Suppose next that the largest prime divisor of n is 7. Again 7^2 cannot be a divisor of n because $\phi(7^2) = 42$. So $n = 7m$, where $\gcd(7, m) = 1$ and

$$\phi(n) = \phi(7m) = \phi(7)\phi(m) = 6\phi(m) = 18.$$

This is impossible because we have seen that $\phi(m)$ is even for all $m > 1$.

The remaining possibilities are $n = 2^r$, $n = 3^s$ and $n = 2^r 3^s$, for positive integers r and s.

For $n = 2^r$,

$$\phi(2^r) = 2^{r-1} \neq 18.$$

For $n = 3^s$,

$$\phi(3^s) = 3^s \times \frac{2}{3} = 2 \times 3^{s-1} = 18,$$

giving the solution $n = 3^3 = 27$.

For $n = 2^r 3^s$,

$$\phi(2^r 3^s) = 2^r 3^s \times \frac{1}{2} \times \frac{2}{3} = 2^r 3^{s-1} = 18,$$

giving the solution $n = 2^1 3^3 = 54$.

So there are four solutions, namely $n = 19, 27, 38$ and 54.

2 By the same argument as employed in Solution 1 above, if $\phi(n) = 14$ then the only possible prime divisors of n are 2 and 3. It follows that n must have one of the prime decompositions $n = 2^r$, $n = 3^s$ or $n = 2^r 3^s$. The corresponding values of the ϕ function are 2^{r-1}, $2 \times 3^{s-1}$ and $2^r 3^{s-1}$. As none of these have a divisor 7 the value 14 cannot arise.

All smaller even numbers arise as values of ϕ:

$$\phi(3) = 2, \ \phi(5) = 4, \ \phi(7) = 6,$$
$$\phi(16) = 8, \ \phi(11) = 10, \ \phi(13) = 12.$$

3 (a) If p and $2p - 1$ are odd primes then, for $n = 2(2p - 1)$,

$$\phi(n) = \phi(2)\phi(2p - 1) = 1 \times (2p - 2),$$
$$\phi(n + 2) = \phi(4p) = \phi(4)\phi(p) = 2(p - 1),$$

and the two are seen to be equal.

(b) If $n = 4p$, where p and $2p + 1$ are odd primes, then

$$\phi(n) + 2 = \phi(4p) + 2 = 2(p - 1) + 2 = 2p,$$
$$\phi(n + 2) = \phi(2(2p + 1)) = \phi(2)\phi(2p + 1) = 2p,$$

and again the two are equal.

4   The key here is the observation that the set of primes which divide n^2 is exactly the same as the set which divide n. Hence

$$\phi(n^2) = n^2 \prod_{p|n}\left(1 - \frac{1}{p}\right) \text{ and } \phi(n) = n\prod_{p|n}\left(1 - \frac{1}{p}\right),$$

from which $\phi(n^2) = n\phi(n)$ follows immediately.

$$\phi(\phi(p^k)) = \phi(p^{k-1}(p-1)) = \phi(p^{k-1})\phi(p-1), \quad \text{as } \gcd(p^{k-1}, p-1) = 1,$$
$$= p^{k-2}(p-1)\phi(p-1) = p^{k-2}\phi((p-1)^2),$$

the final step following from the first part of this question with $n = p - 1$.

Section 5

1   (a) (i)   Since $\phi(17) = 16$ and 6 does not divide 16, there are no integers of order 6 modulo 17.

         (ii) $\phi(18) = 6$ and so there may be integers of order 6 modulo 18. The candidates are those numbers in a reduced set of residues of 18, $\{1, 5, 7, 11, 13, 17\}$. Checking successive powers of these numbers modulo 18 reveals that they have respective orders 1, 6, 3, 6, 3 and 2, the elements having order 6 being 5 and 11.

    (b) $\phi(19) = 18$ and as 6 divides 18 there may be elements of order 6 modulo 19. Given that 2 is a primitive root of 19, Theorem 5.2 tells us that 2^k has order $\dfrac{18}{\gcd(k, 18)}$. This will be equal to 6 when $\gcd(18, k) = 3$, that is, for $k = 3$ and 15. The numbers which have order 6 modulo 19 are therefore $2^3 = 8$ and $2^{15} \equiv 12$.

2   Since $a^3 \equiv 1 \pmod{p}$ we have

$$a^3 - 1 = (a - 1)(1 + a + a^2) \equiv 0 \pmod{p}.$$

Now a is not congruent modulo p to 1 for, if so, a would have order 1. Hence, appealing to Euclid's Lemma, $1 + a + a^2 \equiv 0 \pmod{p}$.

From this congruence we observe that $1 + a \equiv -a^2 \pmod{p}$ and so the successive powers of $1 + a$ modulo p are

$$-a^2, a^4, -a^6, a^8, -a^{10}, a^{12}, \ldots,$$

and, utilizing the fact that $a^3 \equiv 1 \pmod{p}$, these simplify to

$$-a^2, a, -1, a^2, -a, 1, \ldots,$$

showing that $1 + a$ has order 6.

3   If r is a primitive root of the odd prime p then r has order $p - 1$ modulo p and, from Theorem 5.2, the primitive roots of p are the powers r^k for those positive integers k with $\gcd(k, p - 1) = 1$. As $p - 1$ is even we see immediately that k must be odd.

Let s be another primitive root of p. Then $s = r^k$, for some odd k. But now $rs = r^{k+1}$ cannot be a primitive root since $k + 1$ is even.

4   If r is a primitive root of n its powers generate the reduced set of residues $\{r, r^2, \ldots, r^{\phi(n)}\}$. Note in particular that $r^{\phi(n)} \equiv 1 \pmod{n}$, and so from the reduced set of residues we can form pairs $\{r^k, r^{\phi(n)-k}\}$ which have the property $r^k r^{\phi(n)-k} \equiv 1 \pmod{n}$. This pairing accounts for all the integers in the reduced set except two, namely the element $r^{\phi(n)/2}$, which would be paired with itself, and the final element $r^{\phi(n)}$. As $r^{\phi(n)} \equiv 1 \pmod{n}$ it follows that the product of all the elements in the reduced set of residues is congruent to $r^{\phi(n)/2} \pmod{n}$. It remains to observe that $r^{\phi(n)/2} \equiv -1 \pmod{n}$; this holds because $\gcd(n, -1) = 1$ and so -1 must feature in the reduced set of residues, and $(-1)x \equiv 1 \pmod{n}$ has unique solution $x \equiv -1 \pmod{n}$ confirming that -1 is the second element paired with itself.

This argument assumes that $\phi(n)$ is even, which is the case for $n > 2$. The result is trivially true for $n = 2$.

The result is not necessarily true if n does not have a primitive root. For example, consider $n = 8$. The product of the four integers in the reduced set of least positive residues is $1 \times 3 \times 5 \times 7 \equiv 1 \pmod{8}$.

INDEX